MACMILLLAN
ENCYCLOPEDIA
OF THE
ENVIORMENT

Shelved with
Circulating
Encyclopedias

MAC
#4 W9-CQB-394
V. 4

Macmillan Encyclopedia of the Environment

Macmillan Encyclopedia of the
ENVIRONMENT

VOLUME 4

General Editor
Stephen R. Kellert

Associate Editors
Matthew Black

Richard Haley

Macmillan Library Reference USA
New York

Developed, Designed, and Produced by Book Builders Incorporated

Macmillan Library Reference
1633 Broadway, New York, NY 10019-6785

Library of Congress Catalog Card Number: 96-29045

Printed in the United States of America

Library of Congress Cataloging-in-Publication Data

Macmillan encyclopedia of the environment.
     p.   cm.
   "General editor, Stephen R. Kellert"—P. iii.
   Includes bibliographical references and index.
   Summary: Provides basic information about such topics as minerals, energy resources, pollution, soils and erosion, wildlife and extinction, agriculture, the ocean, wilderness, hazardous wastes, population, environmental laws, ecology, and evolution.
   ISBN 0-02-897381-X (set)
   1. Environmental sciences—Dictionaries, Juvenile.
[1. Environmental protection—Dictionaries. 2. Ecology—Dictionaries.]
I. Kellert, Stephen R.   96-29045
GE10.M33 1997   CIP
333.7—dc20   AC

Photo credits are gratefully acknowledged in a special listing in Volume 6, page 102.

M

MacArthur, Robert H. (1929–1974)

▌Ecologist credited with basic discoveries in community ECOLOGY, including the theory of island biogeography and the study of the geographical distribution of plants and animals. MacArthur's early education was in mathematics, but he had an avid interest in natural history. His work on island communities drew early attention.

MacArthur worked closely with Harvard zoologist Edward Osborne WILSON. Together, they examined island animals and developed a mathematical description of the number of SPECIES on islands and also the ways in which island communities develop. Although widely criticized as simplistic, the theory provided a link between observations of the simple nature of island communities and the process of colonization.

MacArthur and Wilson developed mathematical models explaining the relationship between island size and the richness of a community of BIRDS, INSECTS, and other WILDLIFE. This theory has been applied to the design of wildlife nature preserves and wildlife refuges to determine the size of an ENVIRONMENT needed to maintain a particular species stability. Thus, the models help scientists accomplish WILDLIFE MANAGEMENT and protect ENDANGERED SPECIES. [*See also* BIOLOGICAL COMMUNITY; CARRYING CAPACITY; CLIMAX COMMUNITY; ECOSYSTEM; HABITAT; and WILDLIFE CONSERVATION.]

Malnutrition

▌A lack of the proper nutrients needed by the body resulting from improper diet or an inability of the body to absorb the nutrients from food. *Malnutrition,* a term which means "bad nutrition," occurs when people eat foods that do not provide the correct amounts of nutrients to their bodies.

The science of nutrition developed rapidly in the nineteenth century. During this century, scientists identified many of the chemicals contained in food. They also began to discover how these substances were used by the body to maintain health. From such research, the notion of a well-balanced diet emerged. People also began to recognize and understand that a small but well-balanced diet is more healthful than a diet consisting of large amounts of foods that supply nutrients from only one or two food groups.

Nutrients are the substances in foods that the body needs to function properly. Nutrients are classified as proteins, fats and oils, sugars and starches, vitamins, and MINERALS. Some people also consider water a nutrient because it is an essential compound needed for proper health. To maintain proper health, people must eat foods which provide their bodies with all of the identified nutrients in the proper amounts.

A lack of one or more nutrients in the body can result in disease. For example, a disease called *kwashiorkor* is the main dietary disease of young children living in developing countries. Kwashiorkor results from a lack of protein in the diet. Children suffering from this disease often have large, bloated bellies and scrawny, wrinkled, bony limbs. The swollen stomachs of the children do not indicate that they have eaten too many fatty foods. The bloating is caused by an accumulation of liquids in the tissues of the body that results from a lack of protein. People who suffer from kwash-

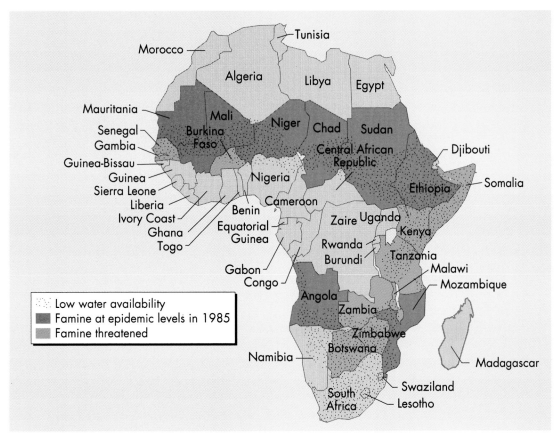

◆ In 1985 famine reached epidemic levels in Africa, where many people continue to suffer from malnutrition caused by famine.

Low water availability
Famine at epidemic levels in 1985
Famine threatened

iorkor live in regions of the world where they either lack access to or cannot afford the meat and dairy products that would provide their bodies with the protein they need.

Many people suffer from diseases and disorders resulting from inadequate amounts of vitamins or minerals in their diet. Beriberi is a vitamin deficiency disease that affects people who do not take in enough vitamin B_1. This disease can result in problems with the nervous system and the heart. Scurvy is a deficiency disease caused by a lack of vitamin C in the diet. This disease causes sore gums and, if left untreated, may lead to a loss of teeth. There are many other deficiency diseases that result from a lack of vitamins and minerals in the body. If treated early, most of the symptoms caused by these diseases can be relieved by adding the missing vitamin or mineral to the diet.

FAMINE, or crop failure, is a major cause of malnutrition. A famine may result from natural events such as devastation of crops caused by FUNGI, INSECTS, or micro-

Common Vitamin and Mineral Deficiency Diseases		
Disease	**Symptoms**	**Missing Vitamin or Mineral**
Goiter	swollen thyroid gland	iodine
Rickets	soft bones and teeth	vitamin D
Anemia	lack of energy resulting from too few red blood cells	iron, vitamins B_6 and B_{12}
Pellagra	diarrhea; rough, scaly skin	vitamin B_3

◆ Some diseases result from a diet that contains too little of a specific vitamin or mineral.

◆ Severe malnutrition caused the deformities in these children.

Mammal

▶ A warm-blooded animal that has hair and feeds its offspring with milk produced in the female's mammary glands. Mammalian brains tend to be large and complex compared with other animals. Humans belong to this class of animals.

Almost all mammals give birth to live young. Mammals are VERTE-BRATES (animals with backbones). Other vertebrate groups include BIRDS, REPTILES, FISH, and AMPHIBIANS.

The four-chambered hearts of mammals help them to maintain nearly a constant body temperature. Animals that can maintain a constant body temperature regardless of the temperature of their surroundings are described as warm-blooded, or endotherms. Hair on the bodies of mammals helps to keep them warm.

KINDS OF MAMMALS

Mammals are extremely diverse—the most diverse of all vertebrate animals. They range in size from pygmy shrews, which are smaller than a human finger, to blue WHALES, the largest animals that have ever lived on Earth. Among the 4,000 living SPECIES of mammals are animals as different as bats, cats, pigs, horses, kangaroos, ELEPHANTS, and humans.

Mammals evolved from reptiles about 220 million years ago. Some mammals still retain traits of mammals of that time. For example, the monotremes, which include only the duckbilled platypus and two species of the spiny anteater, lay

organisms. NATURAL DISASTERS such as droughts and floods may also cause massive crop failure in an area. However, famine is increasingly caused by environmental disturbances resulting from human activities. For example, exploding human populations exhaust once-rich farmland. DEFORESTATION intended to create additional farmland in an area often results in soil EROSION and DESERTIFICATION. These conditions leave the soil unable to support the growth of crops. Thus, as populations grow and the need for food increases, the amount of land available for the growth of crops diminishes.

Malnutrition does not affect only developing nations. In many indus-trialized societies, people are too poor to purchase the foods needed to provide for their families. As a result, diets consist of foods that are inexpensive and often lacking in several of the nutrients needed by the body. People who can afford to buy an adequate amount of food for themselves and their families often eat too much of certain nutrients, especially fats and proteins, leading to obesity, a condition characterized by excessive body fat. It is estimated that as much as 50% of the population of the United States may be overweight. This condition has led to higher occurrences of heart disease, stroke, and other circulatory disorders. [See also HEALTH AND NUTRITION.]

◆ The bat is an example of a flying mammal.

◆ The cougar, also known as a puma, is an endangered mammal.

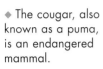

◆ Mammals are the most diverse animals, but they all have hair and feed their young with milk from mammary glands.

eggs and have other reptilian characteristics. Another group of mammals, the marsupials, give birth to tiny underdeveloped young. The young are then suckled in belly pouches. Kangaroos and opossums are examples of marsupials.

Most modern mammals are placental mammals. They carry their young inside the womb until they can survive in the outside world. Ninety-five percent of modern mammals, including humans, are placental mammals.

DIVERSIFICATION, EXTINCTIONS, AND ENDANGERMENT OF MAMMALS

After the extinction of dinosaurs, about 60 million years ago, mammals began to evolve into a great diversity of forms and sizes. Australia and South America, separated from other lands by CONTINENTAL DRIFT, saw the evolution of many endangered species of marsupials. On other continents, placental mammals diversified. Mammals developed many adaptations, such as hard teeth in large HERBIVORES for consuming seeds and stems and ever-growing teeth in rodents for gnawing. One of the groups that evolved a wide variety of species was the primates, a group in which scientists include humans. Over the last 60 million years, most of these mammal species became extinct naturally.

About 2 million years ago, North America was home to thriving populations of horses, elephants, camels, giant ground sloths, and saber-toothed cats. By about 10,000 years ago, all of the mammals in

◆ Orangutans are herbivorous mammals found in Borneo and Sumatra.

that list were extinct. This was just after the time that human beings spread throughout North America. Some scientists think that overhunting by human beings was a chief cause of the mammal EXTINCTIONS.

During the nineteenth century, the bison (also known as the American buffalo) was nearly hunted to extinction. One reason so many bison were killed was to deprive the Sioux, Cheyenne, and other Plains tribes of their primary food source. The bison were killed primarily for their hides, which were sold for as little as fifty cents each. By 1889, only about 550 bison in the entire United States had survived the slaughter.

North American predator populations have suffered, too. Wolves, bears, and big cats have been killed off in many parts of the United States.

In some areas, the lack of large CARNIVORES has contributed to high densities of herbivores like white-tailed deer, which are so numerous in some areas that they can prevent the regrowth of FORESTS. Proposals to reintroduce wolves to some parts of the country have caused heated debate. Many cattle and sheep ranchers oppose plans to reintroduce wolves because wolves do sometimes attack, kill, and eat LIVESTOCK.

North America is not the only place where mammals are in danger of extinction. In Africa, POACHING has brought elephants, rhinoceroses, and GORILLAS to the brink of extinction. Similarly, whaling threatens the existence of the largest of all mammals. [*See also* ENDANGERED SPECIES; ENDANGERED SPECIES ACT; FISH AND WILDLIFE SERVICE; WILDLIFE; and WILDLIFE CONSERVATION.]

◆ Mangrove swamps often support large numbers of birds as well as a great variety of marine animals.

◆ Mangrove forests grow along ocean shorelines.

Mangroves

❙Several SPECIES of trees adapted for growth in the INTERTIDAL ZONE that forms swamps along coastlines in tropical and subtropical areas. Mangroves are quite varied. For example, in the United States, the coast of Florida has red mangroves, black mangroves, white mangroves, and button mangroves. Each type of tree has a slightly different HABITAT.

Despite small differences in habitat, mangroves grow along warm OCEAN shorelines, usually with their bases submerged in the water. Other species of trees cannot survive in the salty, soggy mud of this habitat, usually because their roots die from lack of OXYGEN. Most mangrove species solve this problem by having bases that extend up from the mud into the air.

Mangrove swamps are similar to SALT MARSHES. Both ECOSYSTEMS form in the intertidal zone, an area that is flooded with ocean water at high tide and exposed to the air at low tide. However, salt marshes grow in regions with temperate and cool CLIMATES. In contrast, mangrove swamps form on muddy shores in the TROPICS and subtropics. Mangrove swamps provide habitat for a variety of marine animals. Their roots bring oxygen down into the mud, allowing the mud to support the growth of clams, mussels, oysters, and other animals. In turn, mud and dead plant matter that collect around mangrove roots provide new habitat in which new mangroves can grow. In this way, mangrove swamps grow larger and create even more habitat.

Mangrove swamps also support people. Many of the organisms that live in these swamps are FISH, shrimp, and other shellfish people use as foods. The swamps also protect coastal areas from damage caused by storms, flooding, and EROSION. Mangrove wood is harvested and useful to people because of its hardness. The wood is used for construction and FUEL.

Even though they are one of the richest kinds of WETLANDS, mangrove swamps are frequently destroyed by people. They are often filled in to make land for roads, shoreline homes, businesses, or farms. They may be changed for uses such as growing shrimp in ponds. Mangrove swamps may also be buried in mud that is dredged from harbors to make deep waterways for boats.

During the Vietnam War, many of the mangrove swamps of Vietnam were destroyed through the use of DEFOLIANTS that were used to destroy vegetation in which enemy soldiers could hide. Today, mangrove swamps are still disappearing from Asia, South America, Western Africa, and other regions. In some countries, as much as 75% of the mangroves have been destroyed.

One way to save mangrove swamps is to learn what NATURAL RESOURCES they provide. The swamps can then be managed as valuable resources. A good example of this is occurring in Matang, Malaysia, where mangrove swamps are carefully managed and used. Enough fish and wood products are produced each year to provide one job per 7.5 acres (3 hectares) of swamp. It is estimated that if all the mangrove swamps in Southeast Asia were managed in a way that did not destroy them, they would yield $25 billion per year in products and provide 8 million jobs.

A change from destroying mangroves to conserving them is not easy in most countries. Filling in a swamp may provide larger, quicker rewards than managing it carefully. However, mangrove swamps have an important place in the ECOLOGY of tropical coasts. Thus, it is worth making an effort to conserve them. [*See also* AQUACULTURE; BIOCHEMICAL OXYGEN DEMAND (BOD); BIOGEOCHEMICAL CYCLE; DREDGING; ECOTOURISM; ESTUARY; EVERGLADES NATIONAL PARK; and SUSTAINABLE DEVELOPMENT.]

Marine Mammal Protection Act

◗ A federal law passed in the United States in 1972 to protect and conserve marine MAMMALS. The Marine Mammal Protection Act prohibits the taking or importing of marine mammals in the United States. The act also prohibits the import of products, such as ivory from the tusks of walruses, that are obtained from marine mammals.

Some exceptions to the Marine Mammal Protection Act are allowed if permits are obtained. Permits may be granted for scientific research, public display of marine

◆ The Weddell seal, found in the Antarctic region, is protected under the Marine Mammal Protection Act.

◆ Permits are sometimes granted for the public display of marine mammals, such as the case of Charlotte, a manatee at Sea World in Orlando, Florida.

Marine Pollution

▌A harmful effect on the ocean ENVIRONMENT caused by adding substances to the OCEAN or by changing its temperature. Long before human beings lived on Earth, materials had been carried from the land into the ocean or fallen into the ocean from the ATMOSPHERE. This movement of materials is a natural part of Earth's BIOGEOCHEMICAL CYCLES. Over the ages, oceans have absorbed tremendous amounts of SEDIMENT carried from the land by EROSION, along with huge amounts of PARTICULATES, such as ash and lava from volcanoes. Ocean temperatures have been changed many times by major shifts in the world's CLIMATE.

All these events have changed marine ECOSYSTEMS. Oceans are an enormous natural sink for chemicals and other materials as well as for heat energy. Because of this, people once thought it was impossible to harm the marine environment with POLLUTION. It was once common to dump wastes into the sea freely.

WHY WORRY NOW?

Today, there are many signs that pollutants from human activities are in fact damaging the ocean environment. Near some harbors and river mouths, fish SPECIES die out or show signs of ill health, such as sores and CANCERS. The bodies of some FISH and shellfish contain unusually large levels of HEAVY METALS POISONING or other poisons.

mammals, and accidental capture of these animals during commercial fishing activities. Permits may also be granted to INDIGENOUS PEOPLES who rely on these animals for their survival. For example, the Inuits of Alaska have been granted permits to hunt and use bowhead WHALES. These peoples have relied on bowhead whales for their survival for centuries.

The Marine Mammal Protection Act applies to all whales, DOLPHINS/ PORPOISES, SEALS AND SEA LIONS, walruses, manatees, and sea otters. Violation of the act can result in confiscation of animals, products, and vessels or vehicles. In addition, people violating the act may be subject to fines of up to $250,000. [*See also* ENDANGERED SPECIES; FISHING, COMMERCIAL; INTERNATIONAL CONVENTION FOR THE REGULATION OF WHALING (ICRW); INTERNATIONAL WHALING COMMISSION; and WILDLIFE CONSERVATION.]

Marine PLANTS, which are important in CORAL REEFS and other coastal HABITATS, are disappearing from some areas.

The effects of WATER POLLUTION are usually worst near the coasts and in small closed-in seas, such as the Mediterranean and the Baltic seas. Unfortunately, these are also the areas richest in marine life. Pollution in these regions does not hurt only marine organisms. Worldwide, people get about 14% of the protein they need by fishing the oceans. Some countries, such as Japan, get about 60% of their protein from the sea. Marine pollution is, therefore, a threat to the human food supply.

HOW PEOPLE POLLUTE

There are several reasons why marine pollution has become a problem. Human populations have grown much larger in the last few centuries. As a result, there are more people creating wastes that enter the ocean. The average modern person also uses much more energy and material (which produces more heat and waste) than a person who lived 100 to 200 years ago. People have also invented many new substances, some of which are extremely toxic and make their way into the sea.

Some major human sources of marine pollution are the following:

1. SEWAGE and SLUDGE. In many parts of the world, raw sewage is dumped directly into the ocean. The sludge from SEWAGE TREATMENT PLANTS may also be dumped into the ocean. Each of these activities adds large amounts of nutrients, such as

◆ Rusting cars left on the side of the beach can eventually pollute the ocean.

nitrogen, to marine ecosystems. The increased nutrients can bring about an excess growth of marine plants, including the PHYTOPLANKTON that cause toxic "red tides."

2. Dredge spoil. The bottoms of harbors or waterways are often dredged to keep them deep enough for ships to use. The material, or dredge spoil, removed from the waterways is then dumped into the ocean, where it can damage the HABITAT it falls through as well as the habitat it covers on the ocean bottom.

3. Runoff and river discharge. Water that enters the ocean from rivers or from land runoff often contains SOIL, PESTICIDES, and fertilizers from farmland. Pollutants also wash into the oceans from cities

and from dumps. Much marine pollution comes from these scattered land sources that are hard to identify.

4. Wastes from ocean vessels. A major source of PETROLEUM pollution in marine environments is from "clear out" tanks at sea. This includes sewage, trash, and material flushed out of cargo tanks and ballast tanks. Waste from ships causes a variety of problems. For example, a plastic bag floating in the sea can kill a SEA TURTLE or other animal that swallows it. Small marine organisms present on a ship may be released far from their usual habitat. In their new location these organisms may disrupt local FOOD WEBS.

5. Harbor trash. Like waste from ocean vessels, harbor trash includes

◆ A major source of marine pollution is oil discharged into the ocean by industries.

a variety of items that foul the waters or the bottom of harbors.

6. Spills. OIL SPILLS and other CHEMICAL SPILLS can cause severe damage to ocean ECOSYSTEMS. Some of these materials are directly toxic to organisms and can remain in the environment for years.

7. Waste from industries. Some INDUSTRIAL WASTE, such as FLY ASH from power plants, is sent directly into the ocean. Some RADIOACTIVE WASTE has also been dumped into the ocean. Many toxic chemicals have been released into lakes and rivers. In time, these substances also make their way into the ocean. Other industrial waste enters the ocean by way of the ATMOSPHERE. For example, as FOSSIL FUELS and SOLID WASTE are burned, millions of

tons of pollutants are released into the air. Much of this material later falls into the ocean as solid particles or in rainwater.

8. Heat. Some forms of industry discharge heated water into the ocean. Two examples are NUCLEAR POWER plants and DESALINIZATION plants. On a larger scale, GLOBAL WARMING may cause a widespread rise in ocean temperature. Changes of only a few degrees can cause large changes in some marine ecosystems.

WHAT IS BEING DONE

The oceans are used by many nations. One way to limit pollution is by forming international agreements. Two agreements designed

to reduce marine pollution are the Convention of the Prevention of Marine Pollution by Dumping of Waste and Other Matter (also known as the London Dumping Convention) and the LAW OF THE SEA Convention. Countries can also act alone, trying to keep their citizens and industries from polluting their own coastal waters. For example, the CLEAN WATER ACT of the United States was passed to help reduce pollution of waterways. It is also important to study the way pollutants affect the marine environment and find out where pollutants go after they enter the ocean. [*See also* AGRICULTURAL POLLUTION; ALGAE; ALGAL BLOOM; DREDGING; EFFLUENT; EROSION; EXOTIC SPECIES; EXXON VALDEZ; GARBAGE; HEAVY METALS POISONING; LEACHING; MARINE PROTECTION, RESEARCH, AND SANCTUARIES ACT; MEDICAL WASTE; MINAMATA DISEASE; OCEAN DUMPING; OIL DRILLING; OIL POLLUTION; PHOSPHATES; SALINITY; SEDIMENTATION; and WASTEWATER.]

Marine Protection, Research, and Sanctuaries Act

▶ A U.S. law that controls OCEAN DUMPING, promotes studies on the effects of dumping, and preserves ocean HABITATS. Like the Coastal Zone Management Act of 1972 and several other laws, the Marine Protection, Research, and Sanctuaries Act helps protect the nation's marine ENVIRONMENTS.

◆ The U.S. Coast Guard and the Department of Commerce studied the effects of ocean dumping on marine ecosystems. Their findings helped the EPA to decide the kinds of dumping that should be stopped.

WHY THE LAW WAS PASSED

The Marine Protection, Research, and Sanctuaries Act was passed in 1972 to keep marine NATURAL RESOURCES from being damaged by ocean dumping. In the past, ocean dumping provided a fairly easy and inexpensive way for some cities and industries to get rid of SEWAGE and other wastes. By the early 1970s, several million tons of wastes were dumped each year off the shores of the United States. This was a problem, because dumping creates MARINE POLLUTION.

FISH stocks disappeared from dumping areas. Seafood from those areas often became unsafe to eat. Some beaches and offshore areas also became unsafe for swimming, surfing, boating, or sport fishing. The new law was expected to have more than one good effect. It would not only protect the marine environment from one form of POLLUTION, but it would also help protect commercial fishing businesses and consumers.

WHAT THE LAW REQUIRES

The Marine Protection, Research, and Sanctuaries Act called upon the ARMY CORPS OF ENGINEERS and the U.S. ENVIRONMENTAL PROTECTION AGENCY (EPA) to help control ocean dumping. Before the federal government could give a permit to allow the dumping of waste in the OCEAN, one of these agencies had to study the possible effects of the dumping. The need of a city or a business to dump in the ocean was weighed against the need to keep the environment safe for WILDLIFE and for people who used the ocean for work or recreation. For example, if dumping threatened to decrease the income of fishers, damage a valuable habitat, or make a beach or a bay ugly, unpleasant, or unsafe, the Army Corps of Engineers or the EPA could refuse the request to dump.

In the early 1970s, many effects of ocean dumping were still unknown. The act took a step toward solving that problem. It required the U.S. Department of Commerce and the U.S. Coast Guard to work together and, with other agencies, to study how dumping affected ocean ECOSYSTEMS. What they learned helped the EPA and other agencies decide what kinds of dumping should be stopped. Over the years, EPA regulations have become more strict. New laws to control dumping have been passed. Today, most offshore dumping of pollutants has been banned in the United States.

The act also provides a way to create protected marine areas under the protection of the NATIONAL OCEANIC AND ATMOSPHERIC ADMINISTRATION (NOAA) National Ocean Service. An area may be set aside and protected if it is a good habitat for marine life or a place that people find especially beautiful or good for recreation or research. In the last 20 years, a number of sanctuaries have been established because of this act and other laws that protect the marine environment. [*See also* CLEAN WATER ACT.]

◆ Many brachiopods, which resembled mollusks, became extinct about 350 million years ago and left fossils of their shells embedded in limestone.

Mass Extinction

❚Large-scale disappearance of many SPECIES of organisms within a short period of time. At least five major mass extinctions have occurred during Earth's history. Scientists are not sure about the causes of these mass extinctions. However, evidence suggests that many mass extinctions were caused by severe CLIMATE CHANGES that occurred at different times in Earth's history.

ORDOVICIAN MASS EXTINCTION

The first known mass extinction occurred during the Ordovician period of Earth's history about 435 million years ago. During this time, Earth's shallow, warm OCEANS were teeming with a large diversity of INVERTEBRATES, animals without backbones. Through the study of fossils, scientists have determined that nearly half the species of ocean-dwelling animal families, including trilobites, cephalopods, and crinoids, became extinct at this time. The exact cause of this mass extinction is not yet known, but some scientists believe these animals died out because of sudden changes in ocean levels.

DEVONIAN MASS EXTINCTION

The Devonian period of Earth's history is marked by the EVOLUTION of many species. Also known as the "Age of Fishes," the Devonian period is characterized by the appearance of bony FISH and sharks.

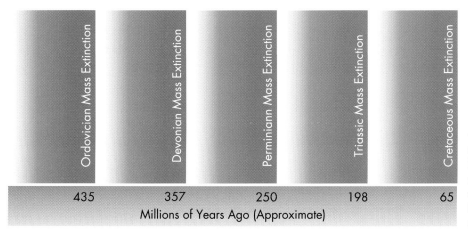

Ordovician Mass Extinction | Devonian Mass Extinction | Perminiann Mass Extinction | Triassic Mass Extinction | Cretaceous Mass Extinction

435   357   250   198   65

Millions of Years Ago (Approximate)

◆ There have been five major mass extinctions during Earth's history. The best known is the mass extinction of the dinosaurs at the end of the Cretaceous period, about 65 million years ago.

◆ This distant relative of the insects became extinct several hundred million years ago.

◆ The early mollusks were forms such as these octopus and squid ancestors.

Marine invertebrates, including many types of arthropods, echinoderms, cnidarians, and mollusks, were also abundant at this time. The appearance of terrestrial plant and animal species are first recorded during the Devonian period. These include several types of simple PLANTS, AMPHIBIANS, and the ancestors of INSECTS.

In the late Devonian period, about 357 million years ago, another large mass extinction occurred. More than one-fifth of all ocean-dwelling families, including corals, fish, trilobites, and brachiopods, died out over a ten-million-year interval.

PERMIAN MASS EXTINCTION

The most severe recorded mass extinction took place at the end of the Permian period about 250 mil-lion years ago. It has been esti-mated that nearly 96% of all plant and animal species became extinct at this time, probably within a period of about one million years. Most marine invertebrate species, including nearly all trilobites, died during this event. Curiously, there was very little EXTINCTION of fish and terrestrial life.

Many scientists suspect that vol-canic eruptions, which sent huge clouds of dust into the ATMOSPHERE, may have caused the great Permian extinction. These scientists believe that dust particles in the air may have blocked out enough sunlight to cause the CLIMATE to become too cold to support many types of life.

TRIASSIC MASS EXTINCTION

During the late Triassic period, approximately 198 million years ago, many types of marine and ter-restrial organisms became extinct. Many species of cephalopods, gas-tropods, brachiopods, and REPTILES were destroyed. Scientists are un-sure about the causes of the Triassic extinction, but evidence points to a drastic cooling of the climate at this time.

CRETACEOUS MASS EXTINCTION

The most recent and best known of all mass extinctions is the event that

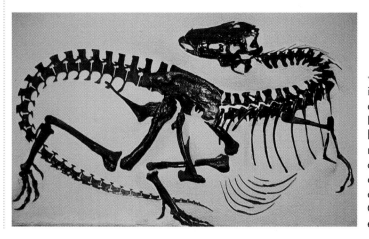

◆ Reptiles evolved into many forms during the Age of Reptiles, which lasted about 160 million years and came to a close 65 million years ago with the Cretaceous mass extinction.

brought about the death of the dinosaurs about 65 million years ago. Although they are the most well-known animals of the period, dinosaurs were not the only organisms to become extinct during the Cretaceous mass extinction. Ammonites, a type of shellfish, and many other types of marine animals perished.

Scientists debate about the cause of the Cretaceous mass extinction. Some scientists suggest that cooling of the climate due to volcanic activity was the primary cause. Others have suggested that many species, including dinosaurs, had been gradually becoming extinct long before the end of the Cretaceous period.

Of the hypotheses suggested for the Cretaceous mass extinction, the "impact hypothesis" proposed by Dr. Walter Alvarez and his colleagues in 1980 is generally most accepted. Alvarez and his co-workers suggest that the mass extinction at the end of the Cretaceous period may have been caused by a large meteorite or asteroid hitting Earth. According to these scientists, the collision sent huge clouds of dust into the atmosphere, effectively blocking out the sun and causing a rapid cooling of the CLIMATE. A large asteroid impact crater from about 65 million years ago has been located off the coast of Mexico's Yucatan Peninsula.

MASS EXTINCTION AND THE LOSS OF BIODIVERSITY

Mass extinctions are not only part of the geological past. Large-scale extinctions have been occurring within the last few thousands of years. Most of these modern extinctions coincide with the appearance of humans. For instance, the disappearance of many large MAMMALS, such as camels, mammoths, mastodons, giant sloths, and lions, from North America seems to have coincided with the appearance of humans on this continent.

Today, human activity is rapidly causing the extinction of many species. Many of these species have not been discovered yet or studied in detail. Some of these species may be potential sources of food or medicines. In addition, some play critically important roles in their ECOSYSTEMS. It is impossible to predict the effects of a loss of BIODIVERSITY. However, most scientists believe that the ecological disturbances that occur when species become extinct can have dramatic consequences for the survival of humans and other species. [*See also* ADAPTATION; EVOLUTION; NATURAL DISASTERS; and VOLCANISM.]

Mass Transit

▶ Forms of transportation that move many people from one place to another at the same time. Mass transit includes buses, trains, subways, airplanes, ferries, and trolleys.

In the United States, most people travel by AUTOMOBILE. Automobiles are a convenient way to move a small number of people from one location to another. However, it is less costly and more energy efficient to move people by mass transit. In addition, the use of mass transit systems involves the use of fewer NATURAL RESOURCES per individual than does the use of private automobiles. Most forms of mass transit are part of public transportation systems run by states or towns.

In many countries, mass transit is subsidized. In the United States, public and private efforts have

◆ Mass transit is a less costly and more energy efficient means of transportation.

more often supported the use of automobiles than mass transit. Only recently has the public begun to support a small amount of mass transit.

Part of the decline of mass transit in the United States resulted from a secret agreement in the 1930s between oil producers and automobile manufacturers. These groups purchased and destroyed trolley cars and trams that once served the people in many cities of the United States. Federal subsidy of gasoline prices and the use of federal funds to build highways have also encouraged the use of privately owned motor vehicles over mass transit.

One benefit of mass transit over private transportation is that mass transit is less costly. For example, in 1993, the average person in the United States spent $1,700 to travel to and from work by automobile in a single year. In Europe, the average person spent less than half that amount using public transportation to travel to and from work.

◆ Each day millions of people use different forms of mass transit to move from one place to another.

The table below shows that people are most likely to use mass transportation to travel to and from work when jobs and the population are tightly clustered. Because mass transit is a less costly and more efficient way to get to work, housing is built and businesses are developed in areas that have city railroad or subway systems. Without mass transit to attract high-density housing, cities of the United States have become more spread out than in other parts of the world. In 1992, more people lived in suburbs than in cities and rural areas combined.

City	Density (cluster of people + jobs per 2.5 acres [1 hectare])	Private Car (% of people using)	Mass Transit (% of people using)	Walking and Cycling (% of people using)
Phoenix, U.S.A.	low	93	3	3
Perth, Australia	low	84	12	4
Toronto, Canada	medium	63	31	6
Hamburg, Germany	medium	44	41	15
Stockholm, Sweden	medium	34	46	20
Vienna, Austria	medium	40	45	15
Tokyo, Japan	high	16	59	25
Hong Kong	high	3	62	35

◆ The greater the density of people and jobs in a city, the more its inhabitants tend to use mass transit rather than private cars.

◆ More people use mass transit in urban areas where business centers are closer to one another.

People living in the United States find cars so convenient that if a city wants to encourage the use of mass transit, it must, at the same time, make cars inconvenient. Many cities are experimenting with ways to stop people from using cars within city limits. For example, in some cities, certain streets are closed to cars but open to delivery trucks, buses, and taxis. This action allows mass transit vehicles to move through the city more quickly than individual cars. In addition, many cities have made fewer parking facilities available to cars or have increased tolls on roads and bridges that provide people access to the city. As the number of cars owned and used by citizens of the United States increases, urban experiments that will ban the use of cars within the city center can be expected in many areas.

People who advocate the use of mass transit argue that it reduces highway congestion, land destruction, AIR POLLUTION, and the dependence of the United States on imported oil. Studies also show that mass transit is an effective way to revitalize urban centers that are in decline by stimulating the development of new businesses in such areas. [*See also* AIR POLLUTION; ENERGY EFFICIENCY; FOSSIL FUELS; and SMOG.]

Mead, Sylvia Earle (1935–)

Marine biologist who studies ocean organisms and has been involved in experiments involving underwater HABITATS for humans. Mead was born in New Jersey in 1935 and was raised in Florida. Throughout her youth, Mead spent much of her time in recreational activities such as swimming and scuba diving in the OCEAN. She later used these recreational activities in her work as a marine biologist.

As a marine biologist, Mead is interested in learning about the BIODIVERSITY of ocean organisms and the habitat of marine organisms. After serving as a fisheries biologist with the U.S. FISH AND WILDLIFE SERVICE, Mead worked as an instructor of college zoology and later as a researcher of marine ENVIRONMENTS.

Among the many studies in which Mead has been involved, one of the most important is considered the *Tektite II* project. In this project, Mead and four other female researchers spent two weeks living in a contained environment below the ocean's surface. The project took place inside a vessel located off the coast of the Great Lameshur Bay. For her work in this study, Mead and her fellow researchers were presented with a Conservation Service Award from the U.S. DEPARTMENT OF THE INTERIOR.

Since her work on the *Tektite II* project, Mead has been involved in a variety of other ocean research projects, frequently as a diver. Through these projects, Mead hopes to learn more about the ocean and the organisms that make it their home. [*See also* COUSTEAU, JACQUES-YVES.]

Medical Waste

Solid and liquid waste from hospitals, clinics, physicians' offices, veterinarians' offices, research centers, funeral homes, and private homes. Medical waste is also called *biohazard waste.*

Medical waste made headlines in 1987 and 1988 when syringes, needles, plastic tubing, glass blood vials, and other health-care items washed up on beaches located along the northeastern coast of the United States. Many people were afraid that diseases would spread from the litter. The beaches were closed to the public. The closures resulted in financial losses to beach communities that depend on the tourist trade.

Investigators found that about 99% of the beach litter came from two sources: medical waste intended for a LANDFILL and trash discarded by users of illegal drugs. The litter was washed ashore partly because of changes in the usual tide and wind patterns.

LOOKING FOR SOLUTIONS

To avoid similar washups in the future, Congress passed the Medical Waste Tracking Act of 1988 to monitor the disposal of medical waste. Under this law, the ENVIRONMENTAL PROTECTION AGENCY (EPA) establishes regulations for the proper disposal of medical waste and tracks the waste. Medical waste from private homes is not covered by the law.

Because of the public outcry about the beach washups in the Northeast, many states passed their own laws to control medical waste. In addition, the Department of Transportation, the Centers for Disease Control (CDC) and the Occupational Safety and Health Administration (OSHA) also became closely involved in the regulation of medical-waste disposal.

DISPOSAL METHODS

Most medical waste is incinerated or burned. Some types of waste are disposed of in landfills (sometimes without disinfection), and some are discharged through sewer systems.

Medical waste can be disinfected by applying dry heat or steam, through CHLORINATION, or with microwave and radio-frequency technology. Electron-beam irradiation and electropyrolysis are now being tested as possible disinfection methods.

About 20% of medical waste contains PLASTIC. Because plastic is inexpensive, it can be thrown away after one use, which helps prevent the transmission of diseases in health-care facilities. However, many plastics are difficult to recycle at present, so methods of decreasing or reusing plastic medical waste are being investigated. [See also HAZARDOUS WASTE; HEALTH AND DISEASE; OCEAN DUMPING; and TOXIC WASTE.]

◆ Medical waste made headlines in the 1980s when syringes washed up on beaches after having been disposed of in the ocean.

Mercury

A liquid, silver-colored heavy metal that is toxic to organisms. Laboratory chemicals, batteries, FUNGICIDES, medications, thermometers, electronic products, and paints may contain mercury or one of its compounds. Improper disposal of any of these products can release mercury into the ENVIRONMENT. Mercury is released into the ATMOSPHERE when PLASTICS are manufactured and FOSSIL FUELS are burned and when cement and steel are produced. Once in the air, mercury can be transported far from its source. Mercury returns to Earth in PRECIPITATION or in particulate form.

It can be carried by RUNOFF and be deposited in aquatic environments.

Once in the environment, mercury may enter FOOD CHAINS. In food chains, mercury increases in concentration in the cells and tissues of BIRDS, FISH, humans, and other organisms through the process of BIOACCUMULATION. Between 1953 and 1960, many people in the Minamata Bay area of Japan became ill or died as a result of eating seafood contaminated with mercury. These people developed a disease called MINAMATA DISEASE. In the 1970s, some TUNA and swordfish sold in the United States contained high amounts of mercury and had to be recalled from stores.

The effects of mercury poisoning include irritability, paralysis, blindness, loss of short-term memory, insomnia, slurred speech, and birth defects. Severe contamination can cause convulsions, coma, and death. Because mercury has been found in plants and animals that people use for food, it is treated as a toxic substance and hazardous waste material. The use of mercury is highly regulated. It is one of the air pollutants for which the ENVIRONMENTAL PROTECTION AGENCY (EPA) sets emission standards. [*See also* CLEAN AIR ACT; CLEAN WATER ACT; HAZARDOUS WASTE; HEAVY METALS POISONING; and TOXIC WASTE.]

◆ Minamata disease, a disease caused by mercury, affected people of all ages, causing such symptoms as paralysis and convulsions, and killing some victims.

Mesosphere

Layer of the ATMOSPHERE, extending from about 30 miles (48 kilometers) to 50 miles (80 kilometers) above Earth's surface. The mesosphere has the coldest atmospheric temperatures—from about 28° F (–2° C) to –216° F (–138° C). The mesosphere also has the highest CLOUDS, which appear in summer just after sunset as silvery strands called *noctilucent*, or nightshining, clouds.

At the uppermost part of the mesosphere is an area known as the *mesopause*, in which the lowest temperature in Earth's atmosphere occurs. Scientists believe the mesosphere has variable air motion like that of the TROPOSPHERE, Earth's lowest atmospheric layer. Scientists point to the zigzag trails of meteors

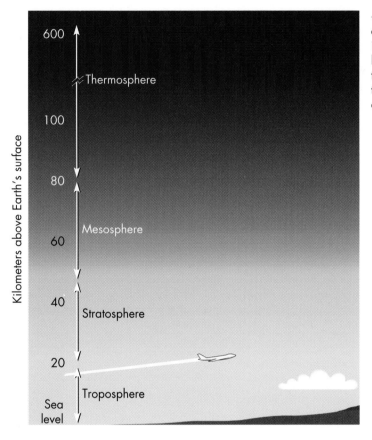

600
Thermosphere
100
80
Mesosphere
60
40
Stratosphere
20
Sea level
Troposphere

Kilometers above Earth's surface

◆ The ever-expanding hole in the ozone layer will affect the mesosphere that lies directly above it.

traveling through the mesosphere as evidence of such air motion.

Immediately above the mesosphere is the thermosphere, Earth's uppermost atmospheric layer. Directly below the mesosphere is the STRATOSPHERE, where a layer of OZONE absorbs the sun's ULTRAVIOLET RADIATION, warming the stratosphere. Scientists believe that the hole in the OZONE LAYER, which is thought to be caused by the use of chlorofluorocarbons (CFCs), has resulted in solar RADIATION reaching Earth. As solar radiation reaches Earth, less warmth is kept in the stratosphere, which disturbs the other atmospheric levels.

Each layer of the atmosphere depends on the others for its stability. Further changes in the stratosphere's ozone layer could harmfully affect the mesosphere and the other layers of Earth's atmosphere. [*See also* GLOBAL WARMING and OZONE HOLE.]

Meteorology

▐▌ The scientific study of Earth's atmospheric conditions that produce WEATHER. Weather refers to what is happening in the ATMOSPHERE in a particular area at a particular time. For instance, *rainy, clear,* and *cloudy* are all descriptions of weather conditions in an area. The object of study in meteorology is the atmosphere, particularly the STRATOSPHERE and the TROPOSPHERE.

MAKING FORECASTS

Meteorologists study the short-term, day-to-day changes that occur in the atmosphere, such as temperature, PRECIPITATION (rain or snow), humidity, air pressure, wind, and cloud cover. They use the information to make forecasts, or predictions, about the weather. To make weather forecasts, meteorologists also use ground-based radar, Earth-orbiting satellites, and weather balloons.

◆ Meteorologists rely on reports from weather stations to make forecasts.

Another branch of meteorology is concerned with the motion of air currents in the atmosphere and the scientific laws associated with the air movement. Research in this field of meteorology involves the use of computers that can predict large-scale weather events, such as hurricanes and tornadoes. [*See also* CLIMATE; CLIMATE CHANGE; THERMODYNAMICS, LAWS OF; ULTRAVIOLET RADIATION; and WATER CYCLE.]

Methane

❿A colorless, odorless gas that is the major ingredient of NATURAL GAS. Sometimes known as marsh gas or swamp gas because it is given off by marshy areas, methane is produced when BACTERIA break down and decompose vegetable matter.

Methane is released into the air during any process that involves the DECOMPOSITION of plant material. For example, rice paddies produce large amounts of methane as old plants decay under water. **Termites** are another major source of methane. When termites eat wood, bacteria in their intestines decompose the wood and release methane gas as a waste product. There are also large quantities of methane locked in the **permafrost** of the frozen TUNDRA in far northern regions.

Human activities also create sources of methane. Scientists estimate that 400–700 tons (360–630 metric tons) of methane are re-leased into the atmosphere every year. Sixty percent of this total results from human activity. LANDFILLS, for instance, are one major source of methane in the atmosphere. When GARBAGE and other wastes decay, huge quantities of methane are produced.

Another human-related source of methane is DEFORESTATION. When trees and other vegetation are cut down to clear land for agriculture, plant materials quickly decompose and release methane into the air. The massive increase in the number of dead trees, in turn, attracts termites, which themselves produce large amounts of methane.

Although methane is produced by natural processes, scientists are concerned about the amount of methane that is released into the atmosphere, particularly because it is a GREENHOUSE GAS—a gas that contributes to the GREENHOUSE

◆ Decomposition of plant material in rice paddies produces large amounts of methane.

EFFECT. Greenhouse gases trap heat near the surface of Earth and prevent its escape into the ATMOSPHERE. Little can be done about methane produced by termites and swamps, but scientists are currently working on technologies that control the release of methane from human sources. [*See also* FOSSIL FUELS and PETROLEUM.]

Mexia, Ynes (1870–1938)

❚ Mexican American explorer and **botanist** who was responsible for the discovery of more than 100,000 SPECIES of PLANTS. Ynes Mexia was born in Washington, DC, and spent much of her early life living in Texas. She lived most of her adult life in Mexico, where she had a business in Mexico City. At the age of 51, Mexia took some courses at the University of California at Berkeley. While there, she developed an interest in the natural sciences—an interest that led her to spend the remainder of her life exploring remote areas to study the BIODIVERSITY of the plant kingdom.

The first exploration for plants conducted by Mexia led her to remote mountain regions on the western coast of Mexico. Many of the regions she studied were inhabited by peoples living in HUNTER-GATHERER SOCIETIES. To conduct her studies, Mexia befriended these peoples to obtain their help as guides through remote regions and also as sources of information about the local names and uses for many plants. Today, this technique is often used by **ethnobotanists** exploring remote rain forest areas to obtain information about the possible medicinal uses of plants.

Mexia's career as an explorer and botanist lasted for 12 years, until her death in 1938. During this time, Mexia explored regions as diverse as the tropical RAIN FORESTS of South America and Alaska's Mount McKinley. In her short, but

---

THE LANGUAGE OF THE ENVIRONMENT

botanist a biologist specializing in the study of plant life.

ethnobotanists plant scientists who specialize in the study of the cultures and languages of peoples native to an area for the purpose of obtaining the help of these people in identifying new plant species as well as their possible uses.

---

productive, career Mexia is reported to have discovered and classified more than 100,000 plants. Some of these plants, such as the *Euphorbia mexiae*—a plant used as a remedy for the bite of a poisonous snake by native peoples in Mexico—were named in her honor. [*See also* MUIR, JOHN.]

Migration

❚ The movement of animals from one location to another, often occurring seasonally and followed by a return to the first location. At some point during an animal's life, its HABITAT may change or the animal's needs may change. An animal may need to travel to a different region to survive those changes. For example, an animal may travel to a new area to stay near sources of food or water, to avoid harsh WEATHER, or to reach a location suitable for mating and for raising its young. The ability of animals to travel, or migrate, allows them to use resources that are not available in only one habitat.

OLD ROUTES AND NEW HAZARDS

Some of the most familiar migrations are the seasonal round-trips made by animals that use different summer and winter habitats. Many migrating bird SPECIES of the Northern Hemisphere, for example, raise their young in northern regions, where summer days are long and

◆ Caribou may encounter oil pipelines in their migration pathways. This will disturb their movements.

◆ Many monarch butterflies migrate more than 2,000 miles from Canada to spend winters in warmer places such as Mexico and Florida.

◆ Snow geese are migratory birds that fly north in spring and south in fall.

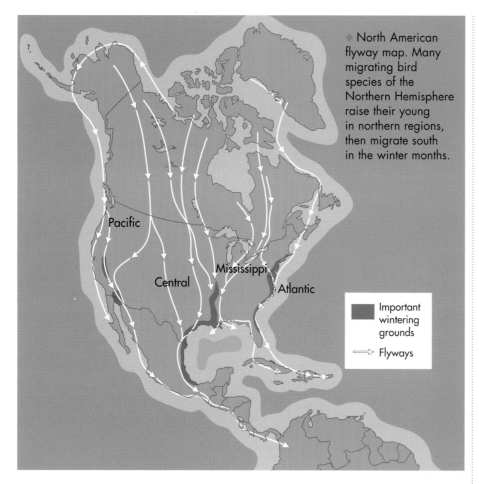

North American flyway map. Many migrating bird species of the Northern Hemisphere raise their young in northern regions, then migrate south in the winter months.

Pacific

Central

Mississippi

Atlantic

Important wintering grounds

⇒ Flyways

return to the ocean to live their adult lives. Eel migration is the opposite of the salmon's. Eels swim thousands of miles from freshwater streams to breed in the ocean.

Although migration can provide a species greater access to resources, the journey is often exhausting and risky. Migrating animals are vulnerable to bad weather. They are also vulnerable to PREDATORS that often gather along migratory routes or at breeding grounds. If the hazards along a species' migratory route become too severe, fewer migrants

THE LANGUAGE OF THE ENVIRONMENT

market hunting the killing of wild animals for the purpose of selling the meat to markets rather than for home use. In the nineteenth century, market hunting of game birds and mammals was a thriving business in the United States and greatly affected wildlife populations.

spawning grounds areas where aquatic animals lay their eggs.

food is abundant. In the fall, these BIRDS travel south to spend the winter months in a milder CLIMATE, where food continues to be plentiful. In mountainous regions, some animals, such as elks and bighorn sheep, spend summers feeding on high slopes. These same animals spend winters in more sheltered foothills and valleys.

Seasonal changes do not affect only temperature. In subtropical regions, the seasonal change is often from wet to dry. Some animals, such as antelopes living in the African GRASSLANDS, migrate to avoid the food and water shortages brought on by droughts.

Many animals must migrate to find suitable spots for reproduction. This is especially true for animals whose young require a habitat that differs from the one that adults prefer. Many frogs and salamanders must migrate to lay eggs in ponds, where their aquatic young can survive. In contrast, pond turtles and sea turtles must leave the water to lay eggs on land. SEALS AND SEA LIONS travel long distances from the open OCEAN to bear their young on beaches, rocks, and ice floes. Adult SALMON swim thousands of miles from their ocean habitat to spawn in freshwater streams. Once developed, the young salmon then

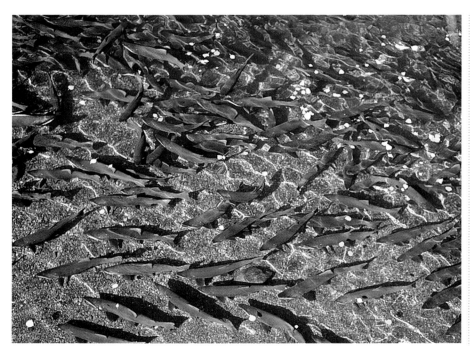

◆ Adult salmon swim thousands of miles from their ocean habitat to spawn in freshwater streams.

may survive the trip or reproduce successfully. This may result in a population decline that places the species at risk of EXTINCTION.

The dangers along the migratory routes of many species have become severe due to human activity. People have always hunted migrating animals. Large-hoofed animals may be hunted and killed at river crossings. FISH may be caught at **spawning grounds**. Migrating birds are often shot or snared while moving in large numbers to new locations. As human technology has improved, direct HUNTING has become more destructive. In the last two centuries in North America, **market hunting** of migratory shorebirds and waterfowl reduced the populations of these birds by millions. For example, HUNTING of the migratory PASSENGER PIGEON (plus other stresses) reduced

this species' population from billions to zero over a few years.

ANIMAL PATHS AND HUMAN BOUNDARIES

Laws created to protect migratory birds in North America have been fairly effective in putting a stop to overhunting, but some problems still exist. Legal protection for migratory animals is often difficult to enact or enforce. The main problem is that animal migration routes differ from human political boundaries. For example, birds that summer in northern Europe may migrate through central and southern Europe and the Middle East before reaching their wintering grounds in Africa. The migrations of wildebeests and ELEPHANTS throughout Africa also cross national boundaries. Monarch butterflies migrate

between the United States and Mexico. Fish, turtles, and WHALES have huge ranges in the open ocean, where no single country controls their fate. Thus, migrating animals often run a gauntlet of different WILDLIFE policies.

Even within a country, policies change abruptly at the boundaries of wildlife preserves and NATIONAL PARKS. Because national parks and preserves are usually too small to contain the entire migration routes of large and mobile animals, migrants often wander out of these parks or preserves into less protected areas.

Another hazard facing migrating animals is the loss or fragmentation of habitat. For example, in addition to nesting and wintering areas, birds need stopover sites in between for resting and feeding. The sites may be used for only a few days or weeks each year. However, if such areas are lost to farming or development, the migratory route can become unusable. As WETLANDS, beaches, and other habitats have been altered, stopover sites have been lost along all the major migration flyways in North America. This has contributed to a decline in many waterfowl and shorebird populations.

Obstacles also interfere with migration. Thousands of migrating birds collide at night with high-rise buildings. SALMON encounter DAMS on their way upstream to spawning grounds. Caribou encounter oil pipelines on their way to seasonal feeding grounds. Similarly, turtles, frogs, and salamanders must cross busy roads that lie between them and their breeding or hibernation areas.

◆ To protect spotted salamanders during their migration to a pond for mating, the residents of Amherst, Massachusetts, close the road where the salamanders travel during this period.

CONDUCTING THE TRAFFIC

There are several reasons for assisting animal migrations instead of obstructing them. Many migratory species are valuable for their beauty. In addition, some species, such as hoofed animals, fish, and waterfowl, also have commercial and recreational value.

Assisting migration takes effort. Migratory routes need to be researched until they are well understood. Laws and treaties must be created and enforced. Habitats need to be preserved or restored. Artificial aids to migration may also need to be used. These aids may include bird-repelling devices on large buildings, FISH LADDERS at dams and well-placed underpasses or tunnels that allow animals to cross pipelines and roads. All these aids are currently being used to some extent. However, much more needs to be done to reduce the disruption of migratory routes. [*See also* ALASKA PIPELINE; FISH AND WILDLIFE SERVICE; HABITAT LOSS; and MIGRATORY BIRD TREATY.]

Migratory Bird Treaty

International agreement established in 1918 between Canada and the United States and later extended to Mexico in 1936 limiting the HUNTING, capture, selling, and killing of migratory BIRDS, such as ducks, geese, and swans. Similar agreements have been made with Japan (1972) and the former Soviet Union (1976).

◆ Canadian geese migrate to the United States and Central America in the fall.

The Migratory Bird Treaty was a response to the overhunting of many bird SPECIES around the beginning of the twentieth century by collectors of feathers and eggs. Conservationists argued in favor of laws protecting migrating birds because many species nest in Canada during the summer and then migrate to the United States and Central America along fixed routes during the fall. The Migratory Bird Treaty provides protection for bird species along their MIGRATION routes, in their breeding territories, and at their wintering areas.

There are exceptions to the Migratory Bird Treaty. Some species of ducks, snipe, and swans can be legally hunted in some states even though they are protected under the treaty. Permits can be obtained to capture some protected species for scientific or educational purposes. [*See also* ANIMAL RIGHTS; ARCTIC NATIONAL WILDLIFE REFUGE (ANWR) AUDUBON, JOHN JAMES; CONVENTION ON INTERNATIONAL TRADE IN ENDANGERED SPECIES OF WILD FLORA AND FAUNA (CITES); ENDANGERED SPECIES; and NATIONAL WILDLIFE REFUGE.]

Minamata Disease

▶The name given to the disease of the nervous system caused by MERCURY poisoning of people living near Japan's Minamata Bay in the 1950s. One thousand people died and many were injured as the result of eating FISH contaminated by mercury. The bodies of the fish absorbed mercury that had been released into the waters of Minamata Bay from the industrial wastes of a factory. People suffering from mercury poisoning absorb the metal into their body through the small intestine. The mercury builds up in the brain and kidneys and causes fatigue, muscle problems, numbness, problems with vision and sometimes mental problems. Mercury poisoning can also cause kidney failure. The chances of complete recovery from mercury poisoning are poor and the condition often results in death. [*See also* CLEAN WATER ACT; FOOD CHAIN; HEAVY METALS POISONING; TOXIC WASTE; and WATER POLLUTION.]

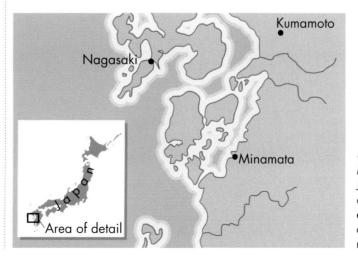

◆ People living near Minamata Bay, Japan, contracted a disease from eating local fish contaminated by mercury.

Mineral Leasing Act

▶ A federal law that authorizes the secretary of the interior to issue leases for the extraction of COAL, NATURAL GAS, PHOSPHATE, sulfur, and other MINERALS from PUBLIC LANDS. The Mineral Leasing Act was passed in 1920. It is regulated by the U.S. DEPARTMENT OF THE INTERIOR.

Since its passage in 1920, additional laws regulating the MINING of minerals from PUBLIC LANDS have been enacted. Such laws required that environmental impact studies of mining on public lands be completed before a lease is issued. These laws include the Mining and Mineral Policy Act of 1970 and the SURFACE MINING CONTROL AND RECOVERY ACT (SMCRA). [*See also* MINING LAW OF 1872; MULTIPLE USE; and RECLAMATION ACT OF 1902.]

Minerals

▶ Inorganic substances occurring naturally in Earth's crust. Minerals are one of Earth's most important NATURAL RESOURCES. There are five categories of minerals:

• Mineral fuels. Mineral fuels include COAL, oil, NATURAL GAS, and radioactive materials such as URANIUM. These fuels are used to make electrical and heat energy and to power engines.

• Ferrous metals. These metals, which are used to make certain kinds of steel, include iron, manganese, chromium, nickel, tungsten, vanadium, molybdenum, and cobalt.

• Nonferrous metals. Nonferrous metals include COPPER, ALUMINUM, LEAD, zinc, tin, MERCURY, and magnesium. These metals are used by manufacturers to make metal products and electrical equipment.

• Precious metals. Precious metals include gold, silver, and platinum. These valuable metals are used to make money, jewelry, tableware, sensitive electrical devices, and dental bridges.

• Nonmetallic minerals. Nonmetallic minerals include SULFUR, quartz, mica, ASBESTOS, and fertilizer minerals. Sulfur is a crucial element in many manufacturing processes. Quartz is used in making glass, watch crystals, computer chips, and for sandblasting. Mica is used in the manufacture of rubber, roofing, paint, electrical equipment, and abrasive materials such as sandpaper. Asbestos was once widely used to make heatproof and insulating materials. However, asbestos is a CARCINOGEN and has been found to cause serious lung disease. Thus, it is no longer used in these products.

Fertilizer minerals are calcium, nitrogen, phosphorus, potassium, and small amounts of trace elements such as magnesium, manganese, boron, iron, copper, sulfur, and zinc. These minerals are used to make fertilizer for farmers and gardeners. [*See also* FOSSIL FUELS; MINERAL LEASING ACT; MINING; OIL DRILLING; PETROLEUM; PHOSPHATE; SMELTER; SOIL; STRIP MINING; SURFACE MINING; and TAILINGS.]

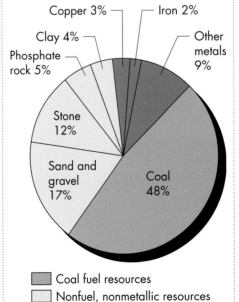

Copper 3% — Iron 2%
Clay 4%
Phosphate rock 5%
Other metals 9%
Stone 12%
Sand and gravel 17%
Coal 48%

☐ Coal fuel resources
☐ Nonfuel, nonmetallic resources
☐ Nonfuel, metallic resources

◆ Almost half of the surface mining of minerals in the United States is mining for coal.

Mining

▶ The extraction of MINERALS from Earth's crust. Rock that contains minerals is called *ore*. Mining began in prehistoric times. However, the INDUSTRIAL REVOLUTION of the nineteenth century enabled people to conduct mining operations on a large scale. At the time of the Industrial Revolution, many parts of the world became dependent upon large quantities of COAL to provide the energy needed to make products in large quantities. Iron also became widely used during this period in history. Since the Industrial Revolution, the increase in human population has made it necessary to make greater amounts of products. The manufacture of products in large numbers has led to an increase in the amounts and types of materials that are mined.

MINING METHODS

All methods of mining involve the removal of SOIL and rock from Earth. The materials may be removed from Earth's surface or from below the ground. The soil and rock over a mineral deposit are called *overburden*. The type of mining that comes to many people's minds are tunnels, often deep and complex, that follow the desired mineral through the overburden. Many mines, however, are not tunnels but open holes.

Three of the most common mining methods are STRIP MINING, OPEN-PIT MINING, and DREDGING. In strip mining, bulldozers clear, or "strip," the overburden from an area to expose the ore. In open-pit mining, a large hole is dug in the ground so that minerals can be extracted.

The world's largest mines are open-pit mines. For example, at the Bingham Canyon open-pit COPPER mine in Utah, more than 3 billion tons (2.7 billion metric tons) of soil and rock have been removed from a 3-square mile (approximatey 7.8 square-kilometer) area to a depth of 2,476 feet (755 meters). This is seven times the amount of material excavated during the building of the Panama Canal.

Dredging is another mining method. Dredging involves the removal of materials from the floor of a body of water such as a lake, river, or ocean. This mining method is sometimes used to obtain sand.

BENEFITS AND DRAWBACKS OF MINING

Without mining, civilization as we know it could not exist. Because it is the way we obtain mineral resources, it is crucial to almost every human endeavor. Without mining, for example, we would not have sources of new metals. However, mining practised without critical safeguards can be dangerous to individuals and to the ENVIRONMENT.

Health Problems

Mining is a difficult and sometimes dangerous business. Cave-ins and poisonous gases in underground mines have killed many miners. Many more miners suffer from illnesses caused by exposure to toxic substances, such as MERCURY or LEAD.

In addition, breathing in particles such as coal dust or ASBESTOS often causes serious and sometimes fatal respiratory diseases, such as black lung disease and lung CANCER.

Habitat Destruction

Mining operations disrupt ECOSYSTEMS. They may also destroy the HABITAT of many SPECIES. For instance, about 350 tons (318 metric tons) of soil, rock, and vegetation must be removed from an area to produce only 1 ton (0.9 metric tons) of copper. This process displaces many organisms from their homes and makes an area vulnerable to soil damage caused by EROSION and POLLUTION.

Pollution

Mining can cause serious POLLUTION problems. Many drainage waters from mining areas have high sulfate

◆ This open-pit mine in Baghdad produces copper.

◆ Waste from mining activities is sometimes disposed of in ponds or lakes and may contaminate the body of water.

Mining Law of 1872

▶ Law that allows any person to lay a mining claim to PUBLIC LANDS and then purchase the land for as little as $5.00 per acre (0.4 hectares). The law was originally enacted by President Ulysses S. Grant to encourage MINING and promote settlement of the West.

The law provides that a person has only to do $100 worth of work on the land and give proof of the discovery of a valuable MINERAL. More than 1.2 million of these claims have been filed for over 25 million acres (10 million hectares) of former public lands. Two thousand of these claims are in NATIONAL PARKS. Large parts of these lands are never mined, but instead are developed and built upon at very large profits for the owners. Claims bought in 1981 near Las Vegas, Nevada, for $1,124 are now worth $2 million. In Colorado, 17,000 acres (6,800 hectares) that contain OIL SHALE were bought for $42,500 and sold a few weeks later for $3.7 million. An Arizona developer receives almost $300,000 a year from a resort hotel built on public land he bought in 1968 for only $155.

The 1872 law contains no environmental protection provisions or land reclamation requirements. For example, if mining occurs on the claimed land, the owner is not required to restore the land that is mined out, a practice known as reclamation.

Though many people call for reform of the Mining Act of 1872, those in the mining industry argue that they need the special provi-

and iron concentrations. When exposed to air, pyrite and marcasite (ferrous sulfide) in coal seams produce sulfuric acid. The iron may also react with SEDIMENTS or mix with alkaline river waters as ferric hydroxide. As a result of chemical reactions with clay and silicates in rocks, water discharged from mines may also contain calcium, magnesium, manganese, and ALUMINUM.

PLANTS can be devastated in areas with WASTEWATER contaminated by copper, nickel, lead, and zinc mines. As water contaminated with such toxic metals soaks into soil, it may also kill the microbes and FUNGI that normally enrich soil through the DECOMPOSITION of organic matter.

Some pollution results from the processes used to refine ores that have been mined. For example, cyanidation is used to separate gold from its ore. The ore is placed in a tank containing a solution of cyanide, a deadly poison. After this treatment, some cyanide remains in the leftover ore.

Lowered Water Tables

Mining can lower the WATER TABLE in a region. Groundwater that collects in the bottom of a mine shaft is pumped to the surface and discharged. In the United States, declines in water levels have resulted in the formation of thousands of sinkholes in Alabama, Georgia, Florida, Tennessee, Missouri, and Pennsylvania since 1900. [See also HEAVY METALS POISONING; LAND USE; LEACHING; MINERAL LEASING ACT; MINING LAW OF 1872; RECLAMATION ACT OF 1902; TAILINGS; and WATER POLLUTION.]

sions of the act to remain profitable. They also maintain that the states have enacted sufficient environmental controls and thus the act need not be amended. According to the mining industry, reforming the act would lead to the loss of American mining jobs and might cause a shortage of affordable minerals and mineral-based products. [*See also* MINERAL LEASING ACT; OPEN-PIT MINING; RECLAMATION ACT OF 1902; STRIP MINING; and TAILINGS.]

Monoculture

▮ Farming practice that involves the cultivation of a single plant SPECIES. To be successful as a farmer, it is necessary to grow crops of high enough quality and in great enough quantity to meet the demands of the market. Many farmers once grew a variety of crops. However, knowing that some types of crops are more profitable than others, many farmers now concentrate their efforts on growing only one or two crops that demand high prices. This practice is known as *monoculture farming*. Monoculture farming can be profitable. However, it creates some environmental problems and may place a crop at greater risk of destruction from INSECTS and plant diseases.

The main environmental problem caused by monoculture farming is depletion of nutrients from the SOIL. Since all the plants grown are identical, they all have the same nutritional requirements. Thus, the plants continually remove the same

◆ Monoculture is practiced in many countries. Women in Madagascar are planting rice.

◆ The growth of only one type of crop on a farm encourages the development of insect populations that feed on the crop and places the crop at great risk of destruction.

nutrients from the soil, and the soil becomes depleted of these nutrients over time. As a result, the soil becomes unable to sustain the crop.

The lack of BIODIVERSITY that results from monoculture farming places plants at a higher risk of destruction from disease and insect pests. Monoculture crops are genetically identical. Thus, all plants grown are susceptible to the same types of diseases. When these plants are grown in great numbers very close together, disease, especially that caused by microorganisms, is able to spread rapidly throughout the population. As a result, an entire crop may be devastated by disease in a short period of time. Similarly, the growth of great numbers of plants creates an almost unlimited food supply for insects that feed upon the plants. With abundant access to food, conditions favor the rapid POPULATION GROWTH of such insects. A greater number of insects, in turn, can damage a greater number of plants, possibly to the point of destroying the entire crop.

Many farmers use chemical substances to combat the problems created by monoculture farming. For example, fertilizers may be added to the soil to replace depleted nutrients. However, such chemicals can be harmful to the ENVIRONMENT, especially if they are carried by RUNOFF or groundwater supplies to lakes and streams. Similarly, the chemical PESTICIDES used to combat disease-causing organisms and insects may also pollute the ENVIRONMENT. In addition to their effects as pollutants, such chemicals may be directly harmful to people working with the crops. If not used carefully, such pesticides may also kill helpful insects or other organisms in addition to their target organism. [*See also* AGRICULTURAL POLLUTION; AGROECOLOGY; BIOACCUMULATION; CARSON, RACHEL LOUISE; DDT; INTEGRATED PEST MANAGEMENT; SUSTAINABLE AGRICULTURE; and TREE FARMING.]

Montreal Protocol

▌A 1987 international treaty that calls for the phasing out of the use of chemicals that break down the OZONE LAYER in the ATMOSPHERE. The

◆ Helicopter installing an airhouse for air conditioning a building. CFCs used in cooling equipment such as air conditioners drift up into the atmosphere, where they break down an important atmospheric layer.

ozone layer, located 15 miles (24 kilometers) above Earth's surface, protects life on Earth from too much ULTRAVIOLET RADIATION, which is produced by the sun. Chemicals called chlorofluorocarbons (CFCs) that break down OZONE have been used in AEROSOL cans and as cooling substances in refrigerators and air conditioners.

The Montreal Protocol on Substances that Deplete the Ozone Layer was signed by representatives of 24 nations on September 26, 1987. It called for the manufacture of CFCs to stay at 1986 levels until 1989. By 1994, manufacturing levels were to drop by 20% and by 1999, levels should decrease by another 30%. Even as the treaty was being signed, many scientists who were detecting 50% decreases in the ozone layer at the South Pole believed a 95% reduction in CFC manufacture should be called for.

In 1989, scientists found ozone decreases over Europe and the United States. Representatives of 120 nations met in London to toughen the terms of the Protocol. In 1990, 93 nations agreed to new restrictions calling for a complete end to CFC manufacture by the year 2000. China and India, which had refused to sign the 1987 agreement, signed the London agreement.

In 1992, as new ozone depletion data was reported, U.S. President George Bush ordered all CFC production in the United States to end by December 31, 1995. In that same year, 87 nations meeting in Copenhagen moved the international phaseout date to 1996. [*See also* GLOBAL 2000 REPORT, THE; GLOBAL WARMING; OZONE; OZONE HOLE; and OZONE POLLUTION.]

Morgan, Ann Haven (1882–1966)

▶American zoologist and educator who worked to have concepts related to ECOLOGY and CONSERVATION placed in the high school science curriculum. Ann Haven Morgan was born in Waterford, Connecticut, in 1882. She received her college education first at Wellesley College and later at Cornell University. Following graduation from college, Morgan worked as an instructor at Mount Holyoke College before returning to Cornell to work toward an advanced degree. While working toward her advanced degree, Morgan developed an interest in aquatic INSECTS—especially mayflies. Her interest in these organisms later led to her being given the nickname "Mayfly Morgan."

Morgan spent much of her career doing research on organisms living in aquatic ECOSYSTEMS. As part of her research, she began to recognize that in addition to studying the organisms, it was necessary to study their HABITATS as well. She emphasized this importance in her book *Field Book of Ponds and Streams: An Introduction to Life in Fresh Water.* The book, which served as a field guide, identified the traits and habitats of a variety of aquatic animal SPECIES. However, it was unique because it also stressed how each group of animals was important within the ecosystem.

Morgan published her last book in 1965. The book, *Kinships of Animals and Man: A Textbook of Animal Biology,* was among the first textbooks to emphasize the need for conservation of nature. Following the book's publication, Morgan's belief in the importance of ecology and conservation became her life's work. She devoted the remainder of her life to having courses of study offered in science classrooms throughout the United States changed to include the ideas of conservation and ecology. To help reach this goal, Morgan spent much time training high school teachers about how to add these topics to their teachings. Her efforts earned her a reputation as a pioneer in the ecological movement of the United States. [*See also* BIOLOGICAL COMMUNITY; BIOME; CARSON, RACHEL LOUISE; and DOUGLAS, MARJORY STONEMAN.]

Muir, John (1838–1914)

▶American naturalist and environmental activist best known for his work that led to the creation in California of Sequoia National Park and YOSEMITE NATIONAL PARK. Muir was born in Dunbar, Scotland, in 1838, and moved as a child to the United States with his family in 1849. While working on a farm in the Wisconsin WILDERNESS during much of his youth, he developed an interest in nature. Years later, Muir attended the University of Wisconsin but left without a degree. In the years that followed, he held several jobs until a factory accident in 1867 left him temporarily blind.

Muir's sight returned a month later. At that time, he was determined both to devote his life to nature and to see as much of North America as possible. He therefore abandoned his job and began a series of trips in the wilderness. The most famous of these was a walk of almost 1,000 miles (1,600 kilometers) from the Midwest to the Gulf of Mexico. During his travels, Muir wrote his thoughts and observations in a journal. His record of his walk from the Midwest to the Gulf of Mexico was the basis of his book *A Thousand Mile Walk to the Gulf,* which was published two years after his death.

When Muir arrived in Yosemite Valley, California, he earned a living by herding sheep and working at various odd jobs. However, he did not permit these activities to interfere with his love of travel and his observations of nature during long hikes. He made many trips, studying WILDLIFE in Nevada, Utah, Oregon, Washington, and Alaska. During his journeys, he became interested in the waterways and spectacular rock formations he saw in Yosemite and elsewhere. Through his observations at Yosemite and the Sierra Nevada mountains in California, Muir became convinced that the unusual landforms he saw there were carved by GLACIERS. In his day, Muir's ideas were controversial, but today most scientists accept these ideas about glacial action.

Muir spent many years in the Yosemite region observing the landscape and studying various ECOSYSTEMS. His observations on the fragility of the ENVIRONMENT led him into a life of activism. As early as 1867, Muir urged the federal government to begin a policy of forest CONSERVATION. He asked the editor of a prominent magazine to come to Yosemite to see the environmental damage caused by the GRAZING of sheep. Later that year, the magazine published several articles alerting the public to the problems at Yosemite and the Sierra Nevada. Due largely to Muir's efforts, the Sequoia and Yosemite NATIONAL PARKS were established by the U.S. Congress in 1890. In 1892, Muir founded the Sierra Club, an organization dedicated to the preservation of WILDERNESS areas.

Muir's environmental activism continued through his later years. In 1897, President Grover Cleveland protected 21 million acres (8.4 million hectares) of unbroken FOREST by creating 13 forest preserves. In spite of complaints from businesses and other commercial interests, Congress upheld that decision. Muir responded by writing two articles that rallied public support in favor of conservation. The following year, Congress approved President Cleveland's national forest protection project.

In 1903, Muir and President Theodore ROOSEVELT embarked on a three-day camping trip to the Yosemite region. Muir convinced the President of the need for conservation throughout the country. Later that same year, Roosevelt added many more acres to the list of federally protected forests. In the first part of this century, Muir lost a debate with his friend Gifford PINCHOT, then head of the U.S. FOREST SERVICE, over the building of a dam in Hetch-Hetchy Valley, then part of Yosemite National Park.

Muir's writings persuaded many Americans to appreciate wilderness as a precious NATURAL RESOURCE. In addition, by personally influencing the highest levels of government, Muir made an invaluable contribution to the cause of conservation. In 1908, the United States government honored Muir by establishing the Muir Woods National Monument in Marin County, California. [*See also* CARSON, RACHEL LOUISE; DOUGLAS, MARJORY STONEMAN; GLACIATION; LEOPOLD, ALDO; MEXIA, YNES; MORGAN, ANNE HAVEN; NATIONAL PARK SERVICE; and PINCHOT, GIFFORD.]

Multiple Use

Ⅾ The management of PUBLIC LANDS for WILDLIFE CONSERVATION, outdoor recreation, and commercial uses such as logging, MINING, and GRAZING. Multiple use applies the belief that NATURAL RESOURCES can have many values. GRASSLANDS, for instance, not only support a variety of WILDLIFE, such as grazing animals, but they also help prevent EROSION by holding SOIL in place and maintaining nutrient levels of the soil. Grasslands can also be used for recreation and scientific research, as well as ranching. Many types of land areas, particularly FORESTS, can have multiple uses.

HISTORY OF MULTIPLE USE

The idea of multiple use emerged in the late 1800s when explorers told the U.S. Congress about the vast untouched forests they had seen in the Wyoming and Montana territories. The explorers suggested that these forests would be severely damaged by logging, mining, and farming if left unprotected. They asked Congress to protect these lands so the public could enjoy them. Congress responded in 1872 by creating the first NATIONAL PARK, YELLOWSTONE NATIONAL PARK.

In 1960, Congress passed the Multiple Use-Sustained Yield Act. This law was designed to prevent the destruction of the NATIONAL FORESTS. The act stated that forests must be managed to "best meet the needs of the American

How well does the use in the first column combine with others?

	Attractive Environment	Recreation Opportunity	Wilderness and Wildlife	Watershed	General Conservation	Wood Production and Harvest
Maintain an attractive environment		Moderate	Good with some limitations	Fully compatible use	Fully compatible use	Limits size of harvests
Recreation opportunity	Good with some limitations		Poor to fair	Low intensity	Moderate	Limited
Wilderness and wildlife	Fully or generally compatible	Poor		Fully	Fully	Generally good, with limitations
Watershed	Fully	Moderate	Good		Fully	Limits size of harvests
General conservation	Fully	Moderate	Fair to good	Fully		Good with limitations
Wood production and harvest	Good with strict limitations	Fair to moderate	Incompatible	Good with limitations on methods	Good with limitations on methods	

◆ The Flathead Indian Reservation in Montana is used for grazing.

people." According to this act, national forests should be used for a balanced combination of "outdoor recreation, range, timber, watershed, and wildlife and fish purposes."

MANAGING FOR MULTIPLE USE

Managing an area for multiple use can sometimes be difficult because many environmental factors may be involved. For example, managing an area for timber removal involves the construction of roads, landings, and other facilities. However, logging disrupts the HABITATS of many SPECIES and restricts the area for human use. Successful multiple use management requires a careful balance between all of the purposes for which an area may be used.

The concept of multiple use has always been controversial. Critics argue that people sometimes have different ideas about what is considered proper use of public land. Thus, multiple use management sometimes causes problems between conservationists and land developers. However, the policy of multiple use is valuable because it prohibits the exclusive use of public lands by any one group in society. Multiple use policy makes sure public lands are used for the greatest number of purposes possible.

MULTIPLE USE OF PUBLIC LANDS

Today, the federal government has set aside about 40% of the land of the United States for use by the public. Most of this land is located west of the Mississippi River and is reserved for use as national parks, wildlife refuges, resource lands, and national forests. Parts of the land are also used as reservations for Native American peoples or as military bases. Most public lands are used for multiple purposes. These purposes may include logging, recreation, WILDLIFE CONSERVATION, mining, and LIVESTOCK grazing.

The DEPARTMENT OF THE INTERIOR and the DEPARTMENT OF AGRICULTURE control most of the public lands in the United States. Several agencies under the direction of these two departments manage the lands. For example, the NATIONAL PARK SERVICE, an agency of the Department of the Interior, manages more than 350 national parks, NATIONAL SEASHORES, monuments, and a variety of historic sites. The FISH AND WILDLIFE SERVICE, also under the direction of the Department of the Interior, manages a system of 478 wildlife refuges. These refuges are present in every state, as well as in five U.S. territories. Resource lands, used mainly for purposes such as livestock grazing and mining, are managed by the BUREAU OF LAND MANAGEMENT, another division of the Department of the Interior. All national forests are managed by the U.S. Forest Service, an agency of the Department of Agriculture. [*See also* ARCTIC NATIONAL WILDLIFE REFUGE (ANWR); BUREAU OF RECLAMATION; DOUGLAS, MARJORY STONEMAN; EVERGLADES NATIONAL PARK; MUIR, JOHN; NATIONAL WILDLIFE REFUGE; SUSTAINABLE DEVELOPMENT; and WETLANDS.]

Mutualism

▶ A close association between SPECIES in which all participating organisms receive benefits. Mutualism is an example of SYMBIOSIS, a close, long-term relationship between species.

The natural world is filled with examples of mutualism. A classic example is seen in termites. Termites are INSECTS that feed exclusively on wood. They can easily harvest wood with their strong jaws. However, termites cannot digest cellulose, a chemical in the cell walls of plant cells. To obtain nutrients from the wood, termites rely on the colonies of BACTERIA that live within their intestines. The bacteria chemically break down the cellulose present in wood, making it possible for the termites to obtain nutrients. In this relationship, the termites benefit by receiving a steady stream of nutrients from the digested wood. The bacteria benefit by being provided with a safe home (the termite intestine) and a supply of food.

Another example of mutualism is the relationship between several species of ants and acacia trees, a family of plants that grows in many parts of the world. The ants that live on acacia trees attack animals that try to land on or eat the tree. The ants also clear away other plants that grow close to the acacia. In these ways, the ants protect the acacia tree. In return, the acacia tree provides nectar and a home for the ants.

Over time, the relationship between the acacia and the ants has developed to the point where neither species can survive and thrive without the other. Such a relationship demonstrates COEVOLUTION, a situation in which species evolve ADAPTATIONS in response to other species. Many symbiotic, or close, relationships between species, including mutualism, have resulted from coevolution. [*See also* ADAPTATION; COMMENSALISM; and PARASITISM.]

◆ Both these ants and the acacia tree on which they live benefit from their close relationship.

◆ A moray eel is kept clean in a mutualistic relationship with the cleaner shrimp, who gets to eat small pieces of marine animals that the eel feeds upon.

N

National Environmental Policy Act (NEPA)

▶ Act passed in 1970 to protect the ENVIRONMENT of the United States. The National Environmental Policy Act (NEPA) requires all federal agencies (except the ENVIRONMENTAL PROTECTION AGENCY [EPA]) to compile information about the impact on the environment of all federal projects that are proposed. This information is to be presented in the form of a Draft Environmental Impact Statement (DEIS). The DEIS is to be made available to all those interested at least 90 days before the project is scheduled to start. Once a DEIS has been made available, the public and all agencies at the city, state, and federal levels can make comments on the DEIS and submit them at hearings held by the agency heading the project. The agency must then revise its DEIS, taking into account comments, objections, and problems raised by the public discussion. The revisions are used to develop the final ENVIRONMENTAL IMPACT STATEMENT (EIS).

A large number of projects affected by the NEPA have been road-building proposals. The public has been responsible for making changes in these projects, such as increasing the landscaping along the highways and creating hiking and biking trails. In New York City, the DEIS was used as a mechanism to stop the creation of Westway, a portion of a federal highway along the west side of Manhattan that borders the Hudson River. A large portion of the new highway would have been built on LANDFILL in the Hudson River. A local consortium of environmental groups showed that the construction would have a negative effect on the striped bass fishery that exists in the Hudson River.

Other projects that fall under the jurisdiction of NEPA include urban planning, water quality, MARINE POLLUTION, sanitation and waste systems, air quality, and AIR POLLUTION control.

OTHER ASPECTS OF NEPA

NEPA set up a three-person COUNCIL ON ENVIRONMENTAL QUALITY (CEQ) to do a number of things, including the creation of guidelines for preparing the DEIS. The CEQ assisted federal agencies in deciding whether they needed to develop a DEIS, coordinated the development of the statement, and collected the DEIS at the proper time. The CEQ also advised the president of the United States on environmental matters, developed new environmental programs and policies, and assisted the president in assessing environmental problems and finding ways to solve them.

The Environmental Protection Agency reviews all DEISs that deal with air pollution, WATER POLLUTION, drinking water supplies, SOLID WASTE, PESTICIDES, RADIATION, and NOISE POLLUTION. In 1993, President Bill Clinton dismantled the CEQ and replaced it with the White House Office on Environmental Policy.

Today, more than 36 states have laws or executive orders requiring environmental impact statements for state projects. Australia, Canada, France, Ireland, New Zealand, Sweden, and several other countries also require environmental impact statements for projects carried out by government agencies. [*See also* ENVIRONMENTAL EDUCATION; ENVIRONMENTAL ETHICS; and ENVIRONMENTAL JUSTICE.]

National Forests

▶ Forest lands administered by the United States government. It has been said that forest land in North America was once so extensive that a squirrel traveling from New England to the Mississippi River would never have had to touch ground along the way. Yet, so many millions of acres of forest land were cleared for agriculture, housing, fuel wood, and wood products, especially during the 1700s and 1800s, that by the late 1800s, much of what had been FOREST was cut. The United States did not adopt a

◆ The old-growth Douglas firs are valuable for their timber and are also used as Christmas trees.

national forest conservation policy until 1891, when a law was passed that authorized the president to designate areas as national reserves. Managing and protecting these reserves, which today are known as national forests, was made the responsibility of the Division of Forestry of the DEPARTMENT OF AGRICULTURE. The Division of Forestry became the Bureau of Forestry in 1901, and, four years later, in 1905, the bureau became the FOREST SERVICE.

As a result of this law and subsequent conservation measures, there are now 156 national forests, and the threat of DEFORESTATION no longer hangs over the United States. In fact, there are more trees growing in the United States now than there were just after World War I.

Wherever forest land has not been used for agriculture or other developments, second-growth forests have taken hold and thrived, and some forests in the eastern part of the United States have gone unharvested since their first cutting many years ago. Some of these are approaching the age and characteristics of OLD-GROWTH FORESTS. Only about 50% of the virgin forests still remains, mainly in the Pacific Northwest.

Approximately two-thirds of the 740 million acres (297 million hectares) of U.S. forests are managed for commercial timber harvest. Most of this managed commercial forest land is privately owned land located in the eastern United States. The rest, located in the West, is administered by the Forest Service, the BUREAU OF LAND MANAGEMENT, and other governmental agencies. In all, about 22% of the nation's commercial forest acreage is found within the boundaries of the national forests.

Congress has mandated management of the national forests for MULTIPLE USE. This means that national forests will be used and maintained in a variety of ways that promotes balance in the use of PUBLIC LANDS. Uses include GRAZING for more than three million cattle and sheep each year; extensive logging and MINING by private businesses; recreation; and protection of WATERSHEDS and WILDLIFE. In practice, these uses are often incompatible for the same portion of forest. Controversy has often broken out among groups promoting different uses.

A particular source of controversy is the way in which the Forest Service manages logging in the national forests. There is considerable incentive, at both the federal and local levels, to encourage timber sales and keep harvests high: 70% of the agency's budget, directly or indirectly, involves timber sales; moreover, counties receive 25% of gross receipts from national forests within their boundaries. Tracts of forest are leased to companies, which harvest the trees and sell the timber. Some critics charge that existing policies have led to aggressive logging of old-growth forests in Oregon and Washington. Critics from some environmental groups and the Government Accounting Office have charged that national forests are often managed in a way that loses money. Generally, of the 120 national forests that are open to logging, fewer than 25 turn a profit. Fees paid for leases do not cover the Forest Service's costs. Some estimates of the Forest Service's total annual losses in its logging operations run as high as $300 million. Some critics estimate that between 1975 and 1993, national forests have lost anywhere from $5.6 billion to $7 billion from below-cost sales. Timber company officials contend that below-cost timber sales keep lumber and paper prices low and provide thousands of jobs.

Not all national forests, however, are managed primarily for timber sales. For example, the Green Mountain National Forest in Vermont has large tracts that are managed with the primary goal of preserving the beautiful views. Often, this may mean cutting a stand of trees in order to open a view to a valley and distant mountains. In Vermont, such views are particularly important during the autumn foliage season. National forests provide a great deal of opportunity for outdoor recreation, with many forests offering extensive opportunities for camping, backpacking, cross-country skiing, HUNTING, and other activities.

The U.S. Forest Service faces a continuing challenge in balancing the needs and desires of a great many users of national forests. Without the national forests and the foresight of people like Gifford PINCHOT and Theodore ROOSEVELT, Americans might have far fewer forests available for a wide range of uses.

In recent years, industry and government foresters, including the Forest Service's professional organization, the Society of American Foresters, have promoted "New Forestry," which emphasizes concern for forests' ecological health and diversity rather than the size of timber harvests. If adopted, the principles of New Forestry may mean significant changes in the way some national forests are managed. [*See also* FOREST PRODUCTS INDUSTRY.]

National Grassland

▶ Any of the GRASSLAND regions protected and administered by the U.S. DEPARTMENT OF THE INTERIOR. Much of the central United States is part of the grassland BIOME. These areas have become important for their natural beauty and for their use in agriculture. For example, many grassland areas are used to grow crops, such as corn, wheat, and oats, that are used by both people and LIVESTOCK for food. In addition, grasslands provide ample space for GRAZING animals, such as cattle, horses, and sheep. Because

◆ About half of the virgin forests in the United States have not been cut down. This one, in the Pacific Northwest, is Siskiyou National Forest in Oregon.

◆ The Buffalo Gap National Grassland in South Dakota is a protected area. Because the midwestern grasslands are so extensive and provide the seeds for many grains, the United States is regarded as the breadbasket of the world.

◆ Livestock grazing is an important use of grassland. Historically, the antelope and buffalo grazed before the West was developed.

the natural BIODIVERSITY—native PLANTS and animal WILDLIFE—are still dominant. The U.S. Department of the Interior maintains these grasslands so the public can experience what the prairies were like before these regions were settled and changed by people.

Among the national grasslands protected by the U.S. government are the Badlands National Monument of South Dakota, Big Bend National Park of Texas, Bowdoin National Wildlife Refuge of Montana, Fort Niobrara National Wildlife Refuge of Nebraska, National Bison Range of Montana, National Elk Refuge of Wyoming, and the Wichita Mountains Wildlife Refuge of Oklahoma. All of these areas are PUBLIC LANDS that are protected and preserved by various agencies within the Department of the Interior. [*See also* LAND STEWARDSHIP; NATIONAL PARKS; NATIONAL PARK SERVICE; and NATIONAL SEASHORE.]

of these uses, no grassland or PRAIRIE in the United States has been left completely untouched by human activity.

Many EXOTIC SPECIES of plants and animals have been introduced to grassland areas. In some cases, these SPECIES have spread throughout the ECOSYSTEM, replacing the NATIVE SPECIES of the area. Other grassland regions have been cleared for the development of housing, agriculture, roads, and industry. Such LAND USES have drastically changed the appearance of the grasslands.

Despite the changes that have occurred in many grassland regions, there are some grassland areas in the United States in which

National Oceanic and Atmospheric Administration (NOAA)

▌An agency of the United States government created in 1970 to manage and monitor ocean resources, provide a variety of oceanic and atmospheric services, and organize scientific research relating to the OCEANS and the ATMOSPHERE. The Natonal Oceanic and Atmospheric

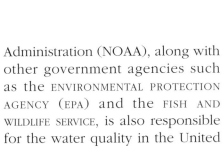

♦ This spacecraft provides the NOAA's National Weather Service with information regarding changes in Earth's atmosphere. This information is used for weather forecasting.

Administration (NOAA), along with other government agencies such as the ENVIRONMENTAL PROTECTION AGENCY (EPA) and the FISH AND WILDLIFE SERVICE, is also responsible for the water quality in the United States.

A division of the Department of Commerce, NOAA has its own major bureaus. These include the NATIONAL WEATHER SERVICE; the National Ocean Service; the National Environmental Satellite, Data and Information Service; the National Marine Fisheries Service; and the Office of Oceanic and Atmospheric Research. One of the most familiar bureaus is the National Weather Service, which monitors WEATHER and alerts people to potential NATURAL DISASTERS like hurricanes, tornadoes, and blizzards.

POSSIBLE CHANGES AT NOAA

In 1995, some members of the U.S. Congress backed a bill that would break apart NOAA and take away some of its environmental responsibilities. The bill, if passed, would transfer administration of the Weather Service and NOAA satellites to the DEPARTMENT OF THE INTERIOR. The bill would also do away with most of NOAA's oceanic and atmospheric research by selling its active research vessels and environmental research labs. This would end NOAA's coastal and WATER POLLUTION studies and threaten the future of NOAA's satellite program, which constantly monitors Earth. [*See also* ACID RAIN; AIR POLLUTION; CLIMATE; GLOBAL WARMING; METEOROLOGY; and OCEAN DUMPING.]

National Parks

Areas set aside by the U.S. government for the preservation of PLANTS, animals, and monuments and for use as natural recreation areas or areas of cultural interest. National parks are established for many reasons. Some are created because they contain abundant WILDLIFE that should be protected for study and enjoyment. Others may contain spectacular or unique geological features. Still others are established for historical purposes.

The United States has one of the most extensive national park systems in the world. Types of areas preserved as national parks in the United States include battlefields and other historic sites, seashores, FORESTS, lakeshores, cemeteries, scenic rivers, places with unique geologic features or of biological interest, and monuments.

HISTORY OF THE NATIONAL PARK SYSTEM

The United States was the first country to establish national parks.

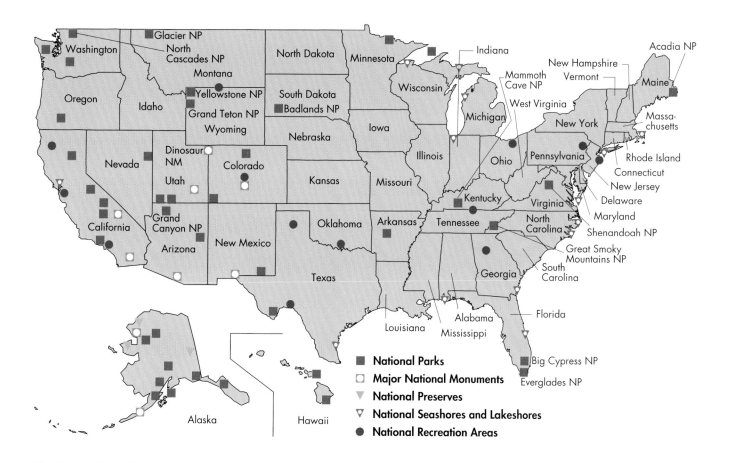

National Parks

- ■ National Parks
- ☐ Major National Monuments
- ▽ National Preserves
- ▽ National Seashores and Lakeshores
- ● National Recreation Areas

◆ The National Park System contains more than 300 parks, seashores, lakes, monuments, cemeteries, and other historic and scenic areas.

Today, all of the areas designated as national parks form the National Park System. The National Park System was established in 1916 with the passage of the National Park Service Act. However, some national parks had existed prior to that time.

The Yosemite Grant of 1864 was the first piece of legislation designed to set aside lands for PUBLIC USE. This act protected 20 square miles (52 square kilometers) of land in California's Yosemite Valley and 4 square miles (10 square kilometers) of giant Sequoia forest. Both areas were placed under the care of the state of California.

In 1872, President Ulysses S. Grant signed into law the establishment of YELLOWSTONE NATIONAL PARK in Wyoming. This became the first national park in the United States. Unlike Yosemite, Yellowstone was to be cared for and managed by the federal government, not by a particular state. Years later, in the 1890s, four more national parks were created. These parks included Sequoia National Park and YOSEMITE NATIONAL PARK.

NATIONAL PARK SERVICE

The early national parks were established for human enjoyment and recreation. Little money was spent by the federal government to care for and maintain these parks. Due to an increasing number of visitors to these areas, some parks suffered from vandalism, littering, OVERGRAZING, and the illegal removal of plants, animals, and other NATURAL RESOURCES.

In 1916, Congress established the NATIONAL PARK SERVICE, an agency within the U.S. DEPARTMENT OF THE INTERIOR. The National Park Service was created to remedy the problems facing the national parks. The mission of the National Park Service is to "conserve the scenery and the natural and historic objects

◆ Mount Rushmore National Memorial is a huge carving in the Black Hills of South Dakota. Presidents Washington, Jefferson, Lincoln, and Theodore Roosevelt are memorialized in figures as high as a five-story building.

VELT reassigned lands that were under control of other government agencies. Two years later, in 1935, the Historic Sites Act gave the Department of the Interior the power to approve additional national historic sites.

National parks are also established through purchases, exchanges, and donations by private citizens. Until 1934, with the establishment of the EVERGLADES NATIONAL PARK, all national parks were created on public land. Everglades National Park became the first national park created on private land. Other national parks developed on private lands include the Great Smoky Mountain National Park in North Carolina and Tennessee and the Shenandoah National Park in Virginia.

NATIONAL PARKS AND RECREATION

Today, more than 275 million people visit national parks in the United States each year. Many people visit these areas to enjoy the unspoiled scenery and wildlife. Others are attracted by recreational activities, such as swimming, camping, FISHING, boating, and hiking. Activities that can damage the parks, such as HUNTING, MINING, GRAZING, OIL DRILLING, and farming, are either prohibited or permitted only in limited amounts. In addition, many people visit the hundreds of national monuments, battlefields, forts, cemeteries, and other historic sites in the national park system to learn more about important people and events in the history of the United States.

and the wildlife therein, and to provide for the enjoyment of the same in such a manner and by such means as will leave them unimpaired for the enjoyment of future generations."

Upon its creation, the National Park Service took over the management of 16 national parks and 21 national monuments. Chicago businessman Stephen T. Mather was the first director of the National Park Service. Mather was committed to the idea of protecting and preserving PUBLIC LANDS. He worked tirelessly to add more areas to the system each year he was in office. When he retired in 1929, the National Park System consisted of 25 national parks, 32 national monuments, and a national memorial.

These regions covered an area of more than 16,000 square miles (41,600 square kilometers).

Today, the National Park System is made up of more than 350 areas that cover an area greater than 125,000 square miles (325,000 square kilometers). National parks are located in all 50 states. Other regions under control of the United States, including American Samoa, the District of Columbia, Guam, Puerto Rico, and the Virgin Islands, also have national parks.

The National Park Service obtains most of its sites when public land is reassigned into the National Park System. For instance, in 1933, 50 new national parks were added to the National Park System when President Franklin Delano ROOSE-

◆ The Grand Canyon is a U.S. national treasure. It is a gash in the earth, in some places 1 mile deep (1.6 kilometers), eroded by the flow of the Colorado River. The exposed rocks are rich in fossils dating back 2 billion years.

◆ Historic monuments, such as the Statue of Liberty, are also designated as national parklands.

The National Park Service employs scientists, historians, educators, and volunteers to care for and maintain the parks and to provide other services to visitors. The mission of the National Park Service has not changed since it was created. However, today, national parks face problems of overcrowding and POLLUTION. The National Park Service has taken many steps to maintain the beauty of the parks while still encouraging recreation. These include tightening controls on water, land, and AIR POLLUTION by limiting the number of people, AUTOMOBILES, and boats that can occupy a park at any one time.

NATIONAL PARKS AROUND THE WORLD

The United States is not the only country that has national parks. Today, more than 1,000 national parks exist worldwide in more than 120 countries. Canada, Great Britain, Japan, Germany, Australia, Brazil, India, Kenya, Tanzania, and many other countries all have national parks.

National parks in other countries face many of the same problems as those in the United States. Many governments provide their park systems enough money to maintain and manage the parks. However, in many places around the world, generous budgets are often not enough. This is particularly true in many of Africa's national parks, such as Serengeti National Park in Tanzania and Tsavo National Park in Kenya, where lions, ELEPHANTS, giraffes, cheetahs, baboons, and other exotic species attract thousands of tourists each year. ECOTOURISM in Africa and other exotic places is big business. Unfortunately, rare species also attract poachers—illegal hunters who sell animals and their products for profit. [*See also* LEOPOLD, ALDO; NATIONAL SEASHORE; NATIONAL WILDLIFE REFUGE; POACHING; SUSTAINABLE DEVELOPMENT; WILDERNESS; WILDERNESS ACT; WILDLIFE CONSERVATION; and WILDLIFE MANAGEMENT.]

National Park Service

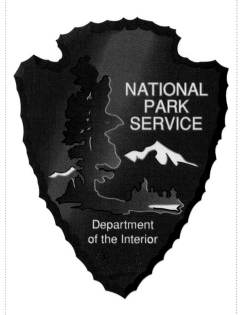

⏹Federal agency that manages the historic and scenic parklands within the National Park System. The National Park Service, a bureau of the DEPARTMENT OF THE INTERIOR, was established in 1916 with the passage of the National Park Service Act. As stated in the law, the mission of the National Park Service is to "conserve the scenery and the natural and historic objects and the wildlife therein and to provide for the enjoyment of the same in such a manner and by such means as will leave them unimpaired for the enjoyment of future generations."

NATIONAL PARKS existed before the National Park Service was created. YELLOWSTONE NATIONAL PARK in Wyoming became the country's first national park in 1872. President Ulysses S. Grant signed a law that placed this scenic WILDERNESS under the management and control of the federal government. Later, Yosemite and Sequoia National Parks were established. These parks were created for human enjoyment, as well as for the protection of WILDLIFE.

After their creation, parks suffered damage from vandalism, illegal harvesting of timber, POACHING, and littering. The National Park Service Act was passed after many years of public concern about the condition of the national parks. People wanted a federal agency that would care for and also manage the parks.

NATIONAL PARK SYSTEM

Today, the National Park Service manages more than 350 areas in the United States that are designated as national parklands. These include parks such as Yellowstone, seashores, wildlife preserves, battlefields, cemeteries, lakeshores, and memorials. The Statue of Liberty and the White House are classified as national parklands.

The National Park System covers more than 80 million acres (32 million hectares) of land in all 50 states. Puerto Rico, the Virgin Islands, Guam, American Samoa, and the District of Columbia also have national parklands.

The National Park Service obtains most of its sites when federal lands are reassigned as national parklands. Other areas are obtained through donations, exchanges, or purchases. Areas may be preserved as national parklands if they have beautiful or unusual natural features, historic value, or attractive recreational features, such as lakes.

◆ An employee of the National Park Service assists visitors at the Channel Islands National Park on Anacapa Island.

Congress must approve all new areas added to the system.

MAINTAINING NATIONAL PARKLANDS

Each year, more than 275 million people visit the national parklands. The National Park Service encourages recreational activities at many of the parklands as long as they do not disturb the surroundings. However, to maintain and preserve the parklands, the National Park Service works hard to keep parklands—and the animals and PLANTS within them—as undisturbed as possible.

The National Park Service employs scientists, historians, educators, and other staff to manage and preserve the features of each parkland. Over the years, they have taken many steps to keep these lands unspoiled, including tightening controls on air, water, and land POLLUTION by limiting the number of cars and boats allowed, as well as by prohibiting destructive activities such as logging, MINING, and HUNTING. FISHING and some LIVESTOCK GRAZING is allowed in national parklands, but in limited amounts. There is a conflict between the original mandate that national parks were

created for recreational use and the present emphasis on the parks' role in preserving NATURAL RESOURCES. This makes managing the National Park Service a difficult job. [*See also* CONSERVATION; ECOTOURISM; EVERGLADES NATIONAL PARK; MUIR, JOHN; NATIONAL SEASHORE; WILD AND SCENIC RIVERS ACT; WILDLIFE MANAGEMENT; and YOSEMITE NATIONAL PARK.]

National Pollutant Discharge Elimination System (NPDES)

▌A system for controlling the quality of POINT SOURCES of WATER POLLUTION. The goal of the CLEAN WATER ACT (formerly known as the Federal Water Pollution Control Act) was to eliminate the discharge of all waste products into streams, rivers, lakes, and other waters. Attaining this goal was quickly viewed as unlikely. However, the goal of decreasing water pollution remains. One way to reach this goal is to issue permits regulating the dumping of waste products into bodies of water.

The U.S. ENVIRONMENTAL PROTECTION AGENCY and environmental quality agencies in each state manage the National Pollutant Discharge Elimination System (NPDES). The NPDES issues permits to regulate the amount of waste released by city wastewater treatment plants, by industries, and by any site where a pipe discharges water to a stream or river. Permits also regulate the release of storm water from parking lots and construction sites. Any company or agency that discharges wastes must obtain a new permit every five years. To obtain a permit, the company or agency must state what kinds of materials (for example, TOXIC WASTES) will be in its WASTEWATER. It must also provide information about the temperature, oxygen content, and acidity or alkalinity of the water. Estimates about the daily volume of waste produced must also be provided.

NPDES permits regulate the amount of waste a discharger can release based on the composition of the wastewater, the amount produced daily, and the minimum flow of the stream or river receiving the waste. National WATER QUALITY STANDARDS for various chemicals are used to estimate if the waste poses a threat to the stream or river that receives the waste. National standards are available for more than 100 priority pollutants. Priority pollutants are those that are common and pose hazards to aquatic and human life.

The NPDES permit process requires a public hearing for dischargers. Once a permit is obtained, the permit requirements may require dischargers to monitor the quality of the wastes discharged. Permit holders are always required to measure concentrations of chemicals known to be in the wastewater. Sometimes permit holders are also required to assess the toxicity of the wastewater. Usually no

toxicity is allowed, because the Clean Water Act states that "no toxic compounds in toxic amounts" are to be discharged into waters of the United States.

Some permits require dischargers to conduct studies of the stream or river that receive their wastes. Data obtained from chemical analysis, toxicity assessment, or field studies must be given to the state environmental agency for review.

National Seashore

▶ A seashore of exceptional beauty, BIODIVERSITY, or importance that is protected and managed by the NATIONAL PARK SERVICE of the U.S. DEPARTMENT OF THE INTERIOR. The Outer Banks, sandy BARRIER ISLANDS set in the water off the coast of North Carolina, became the first U.S. national seashore in 1953. The Outer Banks are now known as the Cape Hatteras National Seashore. When storms bring destructive waves toward the coast, the islands of the Outer Banks absorb much of the impact. In this way, the Outer Banks protect the mainland beaches from the crashing waves. The Outer Banks are crucial to the maintenance of these coastal areas. They are also very beautiful and are home to a diverse community of waterbirds and other beach organisms.

The National Park Service protects and maintains all the National Seashores. Some of the National Seashores include Assateague Island National Seashore in Maryland/Virginia, Fire Island National Seashore in New York, Padre Island National Seashore in Texas, Point Reyes National Seashore in California, and Cape Cod National Seashore in Massachusetts. [*See also* NATIONAL GRASSLANDS and NATIONAL PARKS.]

National Weather Service

▶ The United States WEATHER forecasting bureau. Located in Silver Springs, Maryland, it is a part of the NATIONAL OCEANIC AND ATMOSPHERIC ADMINISTRATION (NOAA) of the Department of Commerce.

Weather has a great impact on many activities on Earth, so forecasting or predicting weather conditions is very important to people. The National Weather Service works to provide timely, precise warnings of severe weather and floods, making use of **weather satellites**, radar, computers, and automated weather data collection stations. Operating at many different sites and using measurements of PRECIPITATION, moisture in the air (humidity), wind direction, wind speed, temperature, and air pressure, **meteorologists** chart their results on computers. They then

◆ Fire Island National Seashore is one of the most spectacular beaches on the mid-Atlantic coast. It is on a barrier island that protects the south shore of Long Island.

◆ Each hour National Weather Service field offices collect weather data such as temperature, barometric pressure, and humidity radioed in from automated weather monitoring stations such as this one.

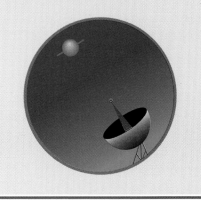

make calculations and forecast the weather. Forecasts focusing on particularly important areas include:

• Ocean and lake weather conditions such as wind speed and tides. Such data are important for shipping, boating, and coastal areas.

• River and flood conditions. Observers from about 6,000 sites provide data that enable meteorologists to forecast river levels and flooding.

• Data about wind and rain. This information is important to farmers. Frost forecasts are also important to farmers who raise certain types of crops, particularly fruit. The forecasts give the farmers time to cover their PLANTS or set up warming devices such as **smudge pots**.

• Fire weather forecasts. Dry conditions in FORESTS and GRASSLANDS make the areas vulnerable to fires because lightning may cause fires in dry foliage. Wind and lightning forecasts are also important elements of fire weather forecasts because high winds will spread the fire.

• Cold waves, ice storms, blizzards, and snowstorm forecasts. These are particularly important to travelers and businesses such as electric and telephone companies whose facilities might be overtaxed or damaged by the effects of the cold weather.

• Severe local storm warnings about dangerous thunderstorms, hailstorms, windstorms, and tornadoes. Such warnings give residents of an area time to prepare properly for such conditions.

• Hurricane warnings. These are important for coastal areas. The high winds, rain, and seas whipped into towering waves by the wind can do a great deal of damage to ships, boats, buildings, highways, utility lines, and beaches in coastal areas. Using planes to measure wind speeds and satellites to track the general path of the hurricane, weather forecasters can usually predict with great accuracy when and where a hurricane will blow onto land areas. This warning system gives authorities time to evacuate people in the storm's path and alert others to prepare by getting emergency supplies and boarding up their windows.

• Weather forecasts. These help pilots plan their flight routes around or over the bad weather. Predictions of snowy or icy conditions also are important to pilots, especially when taking off, landing, or flying at low altitudes.

The National Weather Service also supports research involved in studying the effects of weather on such crops as cotton, pineapple, and corn. Other research involves new methods of forecasting, the effects of sun on weather and climate conditions, measuring and determining the source of energy in severe thunderstorms and tornadoes, and gathering data about the extent of AIR POLLUTION. [*See also* ACID RAIN; CLIMATE; CLOUDS; METEOROLOGY; NATURAL DISASTERS; and WATER CYCLE.]

National Wildlife Refuge

Federally owned tract of land that offers WILDLIFE an area protected from HABITAT destruction. The National Wildlife Refuge System is administered by the U.S. FISH AND WILDLIFE SERVICE. There are more than 400 National Wildlife Refuges in the United States. These lands cover an area of greater than 90 million acres (36 million hectares). Most wildlife refuges exist to provide protected breeding and wintering grounds for migratory waterfowl such as ducks and geese. Other refuges provide protected habitats for threatened and ENDANGERED SPECIES.

HISTORY OF THE NATIONAL WILDLIFE REFUGE SYSTEM

The idea of a national wildlife refuge was introduced in 1903 by President Theodore ROOSEVELT. Roosevelt made Pelican Island in Florida a federal bird reservation. At the time, brown pelicans, along with herons, egrets, and other bird SPECIES, were being overhunted for their valuable feathers. The refuge was aimed at protecting the rare

◆ Hurricanes are powerful storms with winds up to 150 miles per hour (240 kilometers per hour). The National Weather Service collects and disseminates information about these whirling storms so that communities in a storm's path can protect themselves.

◆ Pelican Island, Florida,
established in 1903, was
the first federal wildlife
refuge to protect the
brown pelican.

◆ Musk oxen, shaggy
dwellers of the tundra
and Arctic coastal
regions, were almost
wiped out by hunters in
the mid-1800s. Now
they live in Canada,
Greenland, and Alaska.

and endangered brown pelican, which used the island for breeding. Roosevelt's action prohibited all HUNTING on Pelican Island.

Protection of Migratory Birds

The National Wildlife Refuge System continues to expand. A priority of the U.S. Fish and Wildlife Service is to protect the breeding grounds of migratory BIRDS along their flyways. Thousands of acres of WETLANDS, mostly in the eastern and central United States, have been set aside to protect the breeding and wintering grounds of migratory birds.

Congress passed the Migratory Bird Convention Act in 1929. In 1934, it passed the Migratory Bird Hunting Stamp Act. Both of these acts help provide support to wildlife refuges. The new laws also permit some limited hunting of waterfowl in some wildlife refuges.

Since its creation, the National Wildlife Refuge System has expanded to protect threatened and endangered species. Studies of the habitat needs of birds and MAMMALS have shown that rare and endangered species need protection of their habitats as well as protection from overhunting.

Preserving Habitats of Endangered Species

The emphasis on meeting the habitat needs of animals helped save the whooping crane from EXTINCTION. The whooping crane is a large and beautiful migratory bird. In the 1930s, scientists discovered that the whooping crane population was nearing extinction. The causes of extinction were determined to be

◆ Recreational activities such as fishing are permitted in some wildlife refuges provided those activities do not destroy habitats.

overhunting and HABITAT LOSS. Scientists estimated that there were fewer than 50 whooping cranes living in Canada and the United States at that time. In 1937, the Aransas National Wildlife Refuge was created in Texas to protect the wintering grounds of whooping cranes. Working with the Canadian government, wildlife managers also established protection for the birds on their breeding grounds in northern Canada and along their 2,500-mile (4,000-kilometer) MIGRATION route through North America. Although whooping cranes remain an endangered species throughout North America, their population is slowly growing.

MANAGING WILDLIFE REFUGES FOR MULTIPLE USE

The main purpose of a wildlife refuge is to provide sanctuary for wildlife. However, some refuges in the national system are managed for MULTIPLE USE. Thus, FISHING, GRAZING, farming, hunting, and MINING are permitted in some refuge areas. Such activities, however, are permitted only when they do not interfere with the habitats of threatened and endangered species. While maintaining a protected habitat for animals, wildlife managers must balance the needs of the economy with the needs of the ENVIRONMENT.

Arctic National Wildlife Refuge

The debate over managing wildlife refuges for multiple use became focused in 1980 when Congress passed the Alaska National Interest Conservation Lands Act. This law added 54 million acres (22 million hectares) of Alaskan land to the National Wildlife Refuge System. It also gave the Alaskan government the authority to decide how the lands should be used.

At the center of the debate is the ARCTIC NATIONAL WILDLIFE REFUGE (ANWR) in northern Alaska. This refuge is the second largest in the system. It protects more than 200 different animal SPECIES, including a large number of caribou and millions of migratory birds. The area it protects includes the calving, breeding, and grazing grounds for these animal species.

Geologists believe that several large oil deposits are located beneath the refuge. Since 1980, Congress has considered opening the refuge for oil exploration. However, environmentalists claim that OIL DRILLING in this location is not necessary and would be much too costly for the environment. [*See also* CAPTIVE PROPAGATION; ENVIRONMENTAL ETHICS; MIGRATORY BIRD TREATY; WETLANDS PROTECTION ACT; WILDLIFE CONSERVATION; WILDLIFE MANAGEMENT; and WILDLIFE REHABILITATION.]

Native Americans

See INDIGENOUS PEOPLES

Native Species

▌Type of organism that evolves in association with another SPECIES that is indigenous to an area and thus becomes an integral organism in the functioning of an ECOSYSTEM. The term *native species* is used to identify an organism that is indigenous to a particular region. Thus, it is in direct contrast with nonnative species, or EXOTIC SPECIES, that are frequently introduced to an area through accidental occurrence or the activities of people.

A native species is a natural member of a BIOLOGICAL COMMUNITY. Such a species plays an important role in the functioning of that community. For example, a native species may be a PREDATOR, such as a fox, that keeps populations of other small animals in check. It may be an INSECT that spreads pollen from one PLANT to another. It may be a plant that serves as food for deer or another type of organism. Whatever its role, the native species is an essential member of the community. [*See also* COMPETITION; KEYSTONE SPECIES; and NICHE.]

Natural Disasters

▌Geological or weather-related events, such as earthquakes, volcanic eruptions, hurricanes, floods, and FOREST FIRES, that can devastate human lives, property, and the ENVIRONMENT. Natural disasters can occur at any time and on any continent. Currently, scientists know many details about different types of natural disasters. However, not enough is known for scientists to accurately predict when and where the next natural disaster will occur. For instance, scientists know which environmental conditions are suitable for hurricane formation. They can also track and predict the path of a hurricane fairly accurately once one forms. However, meteorologists, scientists who study Earth's air, WEATHER, and CLIMATE, have not yet discovered why hurricanes do or do not form.

Although the time and place of natural disasters cannot be accurately predicted, knowledge of such events has increased greatly over the past 50 years because of advances in technology. For example, satellites help scientists obtain weather data and other information about Earth. Computers are used to collect and analyze weather data. Other new technologies allow scientists to sense even the tiniest movements within Earth. All these technologies increase our knowledge and help scientists develop methods for protecting against the damaging effects of natural disasters.

EARTHQUAKES

An earthquake is a movement of Earth's crust that causes the ground to vibrate or shake. Earth's crust is made up of many overlapping pieces, called *plates*, that fit together like the pieces of a jigsaw puzzle. The plates move constantly,

◆ Earthquakes, such as this one that jolted Mexico City, cause extensive damage to buildings, roads, and bridges.

rubbing and scraping against each other at their boundaries, or fault lines. Movement at a fault line causes an earthquake. For instance, many of California's earthquakes, such as the 1989 earthquake in San Francisco and the 1994 earthquake in Los Angeles, resulted from movement along the 800-mile-long (1280-kilometer) San Andreas fault. Other fault lines in the United States are located in Hawaii, Missouri, New Jersey, and New England.

Earthquakes can cause extensive damage. Collapsed buildings, roads, bridges, and other structures are common following a strong earthquake. One of the most expensive earthquakes in terms of property damage was the 1989 earthquake in San Francisco. The total repair cost of buildings, roadways, and other structures was near $6 billion. In addition, 62 people died during this natural disaster.

VOLCANIC ERUPTIONS

A volcano is a vent, or opening, in Earth's surface through which melted rock, called *magma,* and gases are expelled. Magma that reaches Earth's surface is called *lava.* Today, there are about 600 active volcanoes around the world. Most of these are located in an area called the "Ring of Fire." The Ring of Fire forms a ringlike pattern of volcanoes in the Pacific Ocean.

Scientists are not able to accurately predict when a volcano will erupt. However, forecasting of volcanic eruptions is generally better than for earthquakes. In 1980, scientists predicted that Mt. Saint Helens, a large volcano in Washington State, would soon erupt. However, they could not pinpoint the exact day when the eruption would occur.

A volcanic eruption can be a dramatic sight, but it can also be very destructive. Effects on humans are limited because few people live near active volcanoes. However, the damage to the ECOSYSTEMS surrounding volcanoes can be extensive. For example, the force from the blast at Mt. Saint Helens flattened vast stretches of CONIFEROUS FORESTS surrounding the volcano. Most living things in this area were killed. In addition, huge clouds of ash and dust were sent into the ATMOSPHERE. Such materials can actually disrupt weather patterns. An eruption of Mt. Pinatubo in the Philippines in 1991 ejected a cloud of dust 4.5 miles (7.2 kilometers) high into the atmosphere, depositing ash thousands of miles away.

HURRICANES

Hurricanes are violent storms that develop in tropical regions. In addition to heavy rainfall, hurricanes have gusting winds of at least 74 miles (120 kilometers) per hour. Hurricanes can develop only over ocean waters having a temperature of at least 80° F (27° C). In addition, these storms develop in areas that are at least 5 degrees latitude north or south of the equator. When these conditions are met, winds begin moving in circular patterns. At first, the wind and precipitation form a tropical storm. When the wind speeds increase to 74 miles (120 kilometers) per hour, they are called hurricanes.

Hurricanes usually move westward at first and then shift toward the north and east. This pattern of movement makes the Gulf of Mexico and many coastal cities of the Atlantic Ocean particularly vulnerable to hurricanes. Hurricanes usually create large waves called *storm surges* that can batter coastal areas.

As a hurricane moves over land, its powerful winds and intense rain destroy homes, businesses, AUTOMO-

BILES, and most other obstacles that get in its way. Fortunately, hurricanes begin to lose energy once they reach land. Often, the hurricanes change into less intense storms over land regions depending upon the weather conditions.

One of the most damaging hurricanes in recent history in the United States was Hurricane Andrew, which battered southern Florida in 1992. When it struck the coastline, this hurricane carried with it wind gusts of up to 160 miles (256 kilometers) per hour. The damage to homes caused by these winds left thousands of people homeless.

FLOODS

Floods are the most common type of natural disaster. No region on Earth is entirely safe from flooding. This includes DESERTS, where the hardened, sun-baked desert floor is prone to flash floods when even small amounts of rain fall.

Floods can be devastating to human lives because cities and towns are often located near bodies of water. Most instances of flooding occur when heavy rains cause rivers and lakes to overflow. Floods cause extensive destruction of ecosystems, automobiles, buildings, and other structures by submerging them under water. Some of the worst floods in recorded history have occurred in recent years. One of the largest floods occurred in China in 1991. According to some estimates, these floods submerged more than 77,000 miles (123,000 kilometers) of land in China. The floods caused the deaths of approximately 2,000 people and left millions of others homeless.

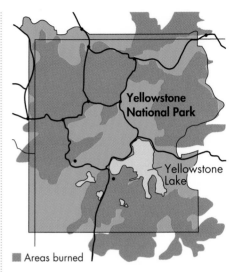

Areas burned

◆ In 1988, nearly 35% of Yellowstone National Park was destroyed by forest fires. Other areas of the United States also experienced wildfires during the drought-filled summer of 1988.

FOREST FIRES

Forest fires usually occur during periods of drought, when conditions are extremely dry. Forest fires can be both helpful and harmful to ecosystems. For example, some tree SPECIES such as the jack pine of Canada and the northern United States must be exposed to fire for their seeds to develop. However, other species of PLANTS and animals that cannot escape fire may be destroyed by forest fires.

Forest fires often burn uncontrollably for weeks. One of the most damaging fires in the United States occurred in the summer of 1988. That year, more than 6 million acres (2.4 million hectares) of forest land burned. One of the most severely damaged areas was YELLOWSTONE NATIONAL PARK in northwestern Wyoming. Fires in this NATIONAL PARK destroyed nearly 1 million acres

(400,000 hectares) of FOREST, 35% of the land, and countless organisms.

Fires can devastate ecosystems. However, ecologists know that fires are common and necessary processes in the functioning of some ecosystems. Small forest fires help remove accumulations of brush and dead wood that might otherwise lead to larger fires. Many animals also depend on forest fires. These animals feed on the sprouts that appear after a fire has cleared the land. [*See also* ABIOTIC FACTORS; AIR POLLUTION; BARRIER ISLANDS; CLIMATE CHANGE; DAMS; DUST BOWL; FAMINE; FIRE ECOLOGY; FLOODPLAIN; PLATE TECTONICS; PRESCRIBED BURN; SUCCESSION; and VOLCANISM.]

Natural Gas

▷A mixture of gaseous HYDROCARBONS, consisting mostly of METHANE, that is obtained from underground reservoirs and is often found near oil deposits. Other hydrocarbons found in natural gas include ethane, propane, and butane. Natural gas, like oil, is a product of the natural DECOMPOSITION of PLANT and animal matter.

EARLY USES OF NATURAL GAS

Although the industrial potential of natural gas began to be realized only during the nineteenth century, ancient Greeks found an interesting nonindustrial use for it. In the belief that natural gas escaping

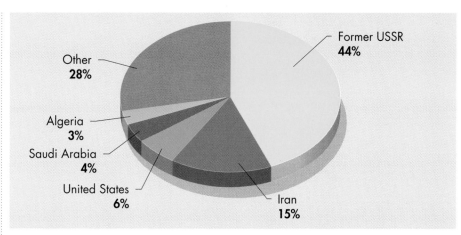

◆ About 59% of the world's natural gas reserves is found in Iran and the former USSR.

from rocks was the breath of the god Apollo, they built a temple at the site, which was named Delphi. A prophetess, or oracle, was kept there, and the god supposedly spoke through the oracle. Other temples were erected over natural gas deposits near Baku, on the Caspian Sea, in about A.D. 600. One temple was still in use in 1880.

Almost 3,000 years ago, however, the Chinese put natural gas to practical use. Gas obtained from deep wells was sent through bamboo pipes. The burning gas was

Natural Gas Composition	
1. methane	(= natural gas)
2. alkanes	butane ethane propane
3. impurities	water sulfur hydrogen sulfide dust

used to boil salt water to obtain salt. The first commercial natural gas well of more recent times was dug in 1821 at Fredonia, New York, which soon after became the first American city to use natural gas for lighting. The gas-lighting industry was shortened due to Thomas A. Edison's invention of the electric light bulb in 1879, but natural gas was used for cooking and heating during the last quarter of the nineteenth century. Gas cooking stoves were exhibited at the Centennial Exposition in Philadelphia in 1876, and the first thermostat-equipped gas stoves appeared in 1915.

MODERN USES OF NATURAL GAS

For many years, there was no practical way to transport natural gas from wells to consumers. Thus, it was considered a waste product of PETROLEUM drilling. So-called casing-head gas was commonly burned at the mouths of oil wells or simply blown into the air to facilitate extraction of oil. By the late 1920s,

World Natural Gas Reserves	
Reserves	(in Billion Cubic Meters*)
World's Total Reserves	86,000
Former Soviet Union	35,000
Iran	11,000
United States	5,800
* 1 billion cubic meters = 1,308 billion cubic yards (approximately)	

welded steel pipes made it practical to carry gas long distances. This increased household and industrial use of natural gas.

By the middle of the twentieth century, natural gas, together with petroleum, was used to produce almost half the valuable organic compounds formerly produced through sugar fermentation. Hydrocarbons derived from natural gas have been used in hundreds of products, including ink, antifreeze, synthetic rubber, PLASTICS, medicines, cosmetics, rayon, and fertilizers. Hydrogen, one of several nonhydrocarbon impurities found in natural gas, is extracted commercially.

DRAWBACKS IN THE USE OF NATURAL GAS

Natural gas burns more cleanly than other FOSSIL FUELS. It emits about half the CARBON DIOXIDE per unit of energy that COAL does and even less per unit than do SYNTHETIC FUELS derived from coal. Unburned natural gas that escapes into the ATMOSPHERE is a GREENHOUSE GAS.

Another problem with natural gas is its scarcity relative to known reserves of other fossil fuels. The world's recoverable reserves of natural gas are an estimated 3,010 trillion cubic feet (86 trillion cubic meters). Significant deposits are located in the former Soviet Union (1,225 trillion cubic feet [35 trillion cubic meters]) and Iran (385 trillion cubic feet [11 trillion cubic meters]). Reserves in the United States amount to 203 trillion cubic feet (5.7 trillion cubic meters). Annual global production, including 16,520 billion cubic feet (468 billion cubic meters) produced in the United States, is about 66.5 trillion cubic feet (1.9 trillion cubic meters). Consumption of natural gas is growing rapidly. However, at the current level of consumption, known reserves will be depleted early in the twenty-first century. [*See also* ALTERNATIVE ENERGY SOURCES; GREENHOUSE EFFECT; and OIL DRILLING.]

Natural History Museum

▶ A museum that houses collections from the fields of **paleontology**, **anthropology**, biology, and **mineralogy**. Many natural history museums also sponsor scientific research in these fields. A natural history museum may contain hundreds of thousands of fossils, arts and crafts, and even live animals and PLANTS.

◆ Natural history museums remind people of their relationship to plants and animals in their environment and the importance of preserving the delicate balance of nature.

◆ The National Museum of Natural History, operated by the Smithsonian Institution, has exhibits and conducts studies on organisms as well as on minerals from Earth and outer space.

well as many fossils of dinosaurs and other prehistoric creatures.

FIELD EXPEDITIONS

Museum scientists often conduct field expeditions to search for new exhibits. They sometimes invite nonscientists to join them. The non-scientists might have the opportunity to work side by side with scientists at an archaeological dig or to observe animal behavior in the wild. [*See also* ECOTOURISM.]

Natural Resources

▶ P roducts of the ENVIRONMENT that humans use. Natural resources are often classified as RENEWABLE RESOURCES and NONRENEWABLE RESOURCES.

Renewable resources include organic and inorganic materials that are replenished through natural cycles. Some organic renewable resources are animals; PLANTS; FUNGI; BACTERIA; and protists, which can be replaced through the process of reproduction. Examples of inorganic renewable resources include SOIL, water, and OXYGEN.

Renewable resources can be exploited and permanently lost if used at a faster rate than they naturally replace themselves. For example, many WILDLIFE SPECIES have

THE LANGUAGE OF THE ENVIRONMENT

anthropology the study of humans and human cultures.

entomologist a scientist who studies insects.

mineralogy the study of minerals.

paleontology the study of past organisms, based on their fossil remains.

Most natural history museums have education departments that work with schools to present programs for students. Students may come to the museum for the programs. Sometimes, museum workers take exhibits out to the schools.

BRINGING NATURE TO LIFE

World-famous **entomologist** Edward Osborne WILSON credits the National Museum of Natural History in Washington, DC, with helping him choose science as a profession. As a young boy, he spent hours at the museum admiring the work of the museum curators. "I could not imagine any activity more elevating than to acquire their kind of knowledge, to be stewards of animals and plants and to put the expertise to public service," Wilson said.

The National Museum of Natural History, part of the Smithsonian Institution, now has more than 30 million specimens in its O. Orkin Insect Zoo. It is the best live INSECT museum in the United States. The museum also has a Hall of Gems, where the Hope diamond and other valuable MINERALS are displayed. In New York City, the American Museum of Natural History contains a reproduction of a RAIN FOREST, as

◆ A "grasshopper" pump is bringing up petroleum in Bakersfield, California.

◆ Renewable products are made from trees in a forest.

become extinct as a result of over-hunting or destruction of HABITAT. Renewable resources have also been degraded and made useless by POLLUTION. For example, the water of the Pigeon River of Cooke County, Tennessee, has become contaminated with a highly toxic chemical called DIOXIN. As a result of this contamination, the water has killed most of the FISH that once lived in it and is unsafe for use by swimmers.

Nonrenewable resources generally cannot be replaced once they are used. Some examples of nonrenewable resources include MINERALS, such as gold, silver, COPPER, and platinum, and FOSSIL FUELS, such as oil, NATURAL GAS, and COAL. Although these materials are produced through natural processes, they are consumed much too rapidly to be replaced at a usable rate.

The exploration and consumption of nonrenewable resources can result in environmental problems. These problems include the production of HAZARDOUS WASTES, AIR POLLUTION, WATER POLLUTION, toxic contamination, and habitat destruction. Recycling can help to conserve many nonrenewable resources, extending their usefulness to people. In addition, some nonrenewable resources are treated to minimize the wastes and pollution associated with their use. [*See also* ALTERNATIVE ENERGY SOURCES; BIOGEOCHEMICAL CYCLE; CHEMICAL CYCLES; CONSERVATION; CONTAINER DEPOSIT LEGISLATION; ENVIRONMENTAL EDUCATION; FRONTIER ETHIC; GARBAGE; GREEN REVOLUTION; HAZARDOUS WASTE MANAGEMENT; LANDFILL; LAND USE; MINING; POLLUTION; RECYCLING, REDUCING, REVISING; and SUSTAINABLE DEVELOPMENT.]

◆ These moths belong to the same species, but in trees with dark bark, natural selection will favor the dark form of the insect. The dark moth is more likely to escape the notice of predators and therefore has a better chance to reproduce.

Natural Selection

▮▮The process in which some members of a SPECIES survive and reproduce in a certain ENVIRONMENT, while others do not. Over time, natural selection may lead to the EVOLUTION of new species.

The idea of natural selection was first developed in detail by Charles DARWIN in his book *On the Origin of Species by Means of Natural Selection* (1859). Darwin suspected that the species present on Earth at any time are not permanent but differ from those existing at other times. He proposed that species evolve or change over time and give rise to new ones. Darwin was not the first to suggest that organisms evolve. However, he was one of the first people to propose that these changes are caused by natural selection. Jean Baptiste de la Marck was a scientist who had suggested similar changes in species.

Darwin pointed out that domestic PLANTS and animals are often

changed using "artificial" selection. In this process, breeders choose a few individuals from a varied group having an unusual feature, perhaps a large size or odd color. Only these individuals are allowed to **breed**. As a result, the next generation has more members with the chosen **trait**. If kept up, artificial selection produces new organisms that are very different from the original group.

In the same way, a wild species can change, depending on which members breed. Some members in a wild species may have some trait that gives them an advantage in surviving and reproducing under certain conditions. For example, an INSECT may be a color that provides especially good camouflage in the extinct. This is why preserving genetic diversity and avoiding unusually rapid change in the environment are important in species CONSERVATION. [*See also* ADAPTATION; BIODIVERSITY; EXTINCTION; and SPECIES DIVERSITY.]

Niche

▌In ECOLOGY, the role an organism or SPECIES plays in its ECOSYSTEM. When describing how a particular species fits into an ecosystem, ecologists often speak of the species' niche. The niche is the species' functional role in the ecosystem. Thus, a species' niche is all its interactions with NATURAL RESOURCES.

To illustrate niche, consider the American alligator. The alligator is a CARNIVORE—a meat-eating organism. The alligator is most active when it is warm. Alligators ensure their supply of fresh water by enlarging mud-lined depressions that fill with rainwater, forming depressions called "wallows" or "gator holes." The wallows attract frogs, which lay their eggs in the fresh water, and herons, which feed on frogs and FISH and nest in surrounding trees. The eggs and babies of BIRDS that fall out of the trees provide food for the alligators and for several species of snakes. Every interaction of the alligator with its ENVIRONMENT is part of its niche. These interactions determine where the alligator can live and what other organisms can **coexist** with it.

◆ Penguins are aquatic birds of the Southern Hemisphere. Millions of years ago they lost their ability to fly, their wings developing into flippers for swimming.

place where it hatched. The camouflage may allow the insect to be one of the few in its generation to escape PREDATORS long enough to mate and have young. As a result, the next generation of insects may have more individuals with gene(s) for this color. If this process of natural selection continues for many generations, the whole population may acquire the new trait and thus become better adapted to its HABITAT. If, over time, natural selection causes enough change in the population, the population may evolve into a new species.

Though natural selection allows populations to adapt to local conditions, it has limits. Natural selection works only if members of the population are genetically different from one another. If all organisms are genetically identical there can be no selection of individuals whose genes are better suited to certain conditions. Thus, variation and GENETIC DIVERSITY are needed for natural selection to occur. Similarly, the more variation present in a population, the more likely it is that natural selection will occur.

Another limit on natural selection is the time it takes. A trait favored by selection takes at least one generation, and often more, to become common in the population. If a habitat changes faster than a species produces new generations, natural selection may allow for EXTINCTION of a species. Thus, natural selection may also place a species at a higher risk of becoming

◆ An aspect of the woodpeckers' niche includes drilling holes in tree trunks to find insects, which they eat. The red-headed woodpecker (right) and the downy woodpecker (left) are two of the more than 23 species that live in North America.

HUTCHINSON'S DESCRIPTION OF NICHE

Ecologist G. Evelyn Hutchinson has suggested that an organism's niche be thought of as a space with many dimensions. Each dimension is a measure of the species' interaction with its environment. The niches of all the species that live together in an ecosystem can be thought of as balloons of various shapes and sizes that are packed into a box. The box with its balloons represents the BIOLOGICAL COMMUNITY in the ecosystem.

Consider a study of the blue-gray gnatcatcher, a small bird that forages for INSECTS in California woodlands. Using Hutchinson's view, two "dimensions" of the gnatcatcher's niche were measured—the lengths of insects caught by the birds and the heights above the ground at which the birds searched for food. When plotted together on a graph, these measurements revealed the birds' activities in two dimensions. The graph showed that insects were most often captured 3.27 to 5.45 yards (3 to 5 meters) above the ground and were most likely to be 0.156 inches (4 millime-

ters) long. A third dimension could be added to this graph, representing perhaps the time of day when the birds looked for food. Such data would produce a three-dimensional graph. Any organism's niche has many more than three components.

◆ Because they are not a keystone species, removing pocket mice from their desert habitat made no difference to the ecosystem.

◆ Alligators are keystone species. They dig deep depressions, or "gator holes," which collect fresh water that provides a sanctuary for aquatic life as well as food and water for animals during dry spells.

◆ Kangaroo rats are also keystone species. Their removal changed the shrubland within 12 years, because the rats ate primarily the large seeds of grasses. When they were gone, the grasses flourished.

These cannot all be drawn on a graph, but they can all be described mathematically.

The fundamental niche of a species represents the range of conditions under which it can survive. In practice, different aspects of the environment may restrict a species to a narrower, realized niche. For example, COMPETITION with humans now restricts the niche of most species. Consider again the niche of the American alligator. The realized niche of alligators in the twentieth century has been increasingly reduced by competition with humans for living space and fresh water.

KEYSTONE SPECIES

Some species, called KEYSTONE SPECIES, are central to a biological community's organization. The niches of a keystone species actually create niches for many other species.

Some ecosystems are obviously organized around keystone species. For instance, a SALT MARSH consists of cordgrass (*Spartina*) and the species that live on it. If the cordgrass were removed, we could no longer recognize the ecosystem as a salt marsh. Thus, cordgrass is an example of a keystone species. Similarly, a few species of decidu-

ous trees provide niches for dozens of other species in a DECIDUOUS FOREST. Such trees are therefore essential to the existence of the FOREST.

In each of the above examples, it is the PLANTS that are essential to the ecosystem. It is less obvious that animals such as alligators at higher TROPHIC LEVELS can act as keystone species. However, experiments conducted by scientists in New Mexico showed that they can. The scientists removed various species, such as all ants, all **rodents**, and various individual species of ants and rodents, from fenced plots. They found that removing most of the species had no noticeable effect. However, removing three species of kangaroo rat changed the ecosystem from shrubland to GRASSLAND within 12 years.

The changes that occurred when the rats were removed showed that these rodents have two main effects on the ecosystem. First, the rats feed primarily on large seeds. Second, the rats disturb the SOIL by digging for seeds. Thus, when the rats were removed from the ecosystem, tall grasses with large seeds established themselves. These new grass species supported specialized rodents that had never before been seen in the area. In this case, the kangaroo rats' niche determined what niches remained in the HABITAT for other species. [*See also* ADAPTATION; COEVOLUTION; EVOLUTION; and SYMBIOSIS.]

NIMBY

▌▶The first letters of the phrase "Not In My Backyard," which reflects an attitude that opposes operating some kinds of technological projects near homes and communities. Such projects include RADIOACTIVE WASTE dumps, incinerators, airports, factories, power lines, oil refineries, and laboratories. NIMBY opposition can also be turned toward a new shopping mall, a certain kind of restaurant, or a video game entertainment center, if residents fear that such businesses might cause property values to fall or attract criminals.

NIMBYism has its roots in a basic distrust of science, technology, and the business community. In the late 1950s, science and technology were treated as a kind of magic that provided better medical care, a better way of life, and great wealth. People believed science could do anything. Then, in the 1960s, Rachel CARSON's book, *Silent Spring,* pointed out the environmental dangers of dichlorodiphenyl trichloroethane (DDT) and other PESTICIDES. Carson wrote that scientific and technological organizations were blind to the social and environmental effects of their work and products.

Since the 1970s, a string of environmental disasters has fed the public's mistrust of science, technology, and big business. In 1976, in Milan, Italy, a chemical factory explosion covered 4,000 acres (1,600 hectares) of land with a dangerous chemical. Scores of animals died and many people became extremely sick. In 1978, the residents of LOVE CANAL, New York, discovered that the constant illnesses they suffered were caused by the poisons in the chemical waste dump on which their town was built. Pennsylvania's THREE MILE ISLAND nuclear power plant had a serious accident that threatened

◆ NIMBY demonstrators succeeded in shutting down a government nuclear weapons plant in Colorado in 1989, eventually igniting the entire controversy about nuclear waste storage sites.

◆ Very strong feelings were aroused at this Washington, DC, protest (left) as well as in Liverpool, Ohio (right), when residents felt threatened by a proposal to build waste incinerators in their area.

to release RADIATION over a large region. More disasters followed.

By 1985, a poll revealed that more than half of Americans felt that they had little control over the risks of daily life. Many said they believed that technology was out of control. From this perception comes fear. Such fear has grown within communities. When a new technological project is announced in a community, often NIMBYs organize, hold meetings, write pamphlets, make signs, and defeat or stall the project.

In some cases, NIMBY actions are selfish, reflecting a community's lack of willingness to shoulder a burden. In Washington, DC, NIMBYs defeated a plan to dump or burn SLUDGE left over from cleaning WASTEWATER. Then they rejected a plan to turn the sludge into fertilizer. The WATER TREATMENT authorities were left wondering what to do with the sludge removed from the wastewater produced by the very NIMBYs who had opposed every disposal plan that was proposed.

Ray Brady of the Association of Bay Area (San Francisco) Governments said, "NIMBYism looks at me, myself, and mine. We're losing a sense of community responsibility. There are certain things we generate—waste, traffic, and so on—and we have a collective responsibility to deal with these things."

WE ARE THE EXPERTS

NIMBYism is fueled by the attitudes of many representatives of science, technology, and business. They may fail to be open about the details of their project, keeping many of their plans secret. They often display little interest in the concerns of the public. When asked for details, their response may be "Trust us. We are the experts." They may not discuss risks; if they do, they may call them very small. They may offer little proof for their claims about risks. Project managers may use the absence of or the vague wording of regulations as a reason for ignoring certain hazards.

Such attitudes produce distrust of scientific information, business, and government regulations, as well as fear of science and technology in many members of a community. Sometimes, NIMBYs search for a bit of scientific data that does not match the data of the project managers. Using this data, they may create a horrific scenario of hazards that is unlikely.

For example, in a campaign opposing the construction of a university biological laboratory in a residential area in California, a NIMBY flier called the project "a special danger to firemen, police, children, and YOU! Explosions! Leaks! Toxic spills!" The flier also distorted the size of a tower that would be part of a monitoring system to check for emissions from the laboratory. NIM-BYs sometimes use scenarios such as this to fuel the fears of fellow citizens.

To their credit, NIMBYs have brought up many health and environmental issues that science and technology representatives have ignored or were honestly unaware of. In some cases, NIMBYs have forced businesses to find alternative ways to achieve certain goals. In this respect, NIMBYs are calling on businesses to account for their actions. [*See also* ENVIRONMENTAL EDUCATION; ENVIRONMENTAL ETHICS; ENVIRONMENTAL IMPACT STATEMENT; and ENVIRONMENTAL JUSTICE.]

Nitrogen Cycle

▶The circulation of nitrogen through the nonliving environment and living organisms. All living things require nitrogen to survive. It is a building block of proteins, amino acids, DNA, and other important organic compounds. Nitrogen is one of the most common elements in Earth's surface—78% of the ATMOSPHERE is composed of nitrogen gas, or N_2. But the vast majority of living organisms cannot make use of this form of nitrogen. PLANTS, animals, FUNGI, and most other organisms can only absorb nitrogen in organic compounds such as ammonia (NH_4), nitrite (NO_2), or nitrate (NO_3). Nitrogen that is put into this form is called *fixed nitrogen*, and NITROGEN FIXING happens by several means.

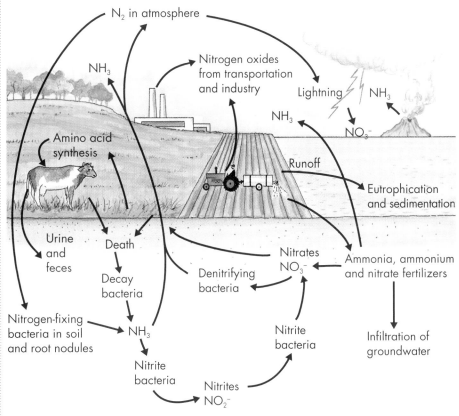

◆ The continual movement of nitrogen from organism to organism and between organisms and the physical environment is shown here in the nitrogen cycle.

Nitrogen Fixing

Nitrogen is fixed into organic compounds by lightning, volcanic eruptions, the burning of FOSSIL FUELS, and by the actions of particular SPECIES of BACTERIA and blue-green ALGAE known as nitrogen-fixing organisms. Some of these bacteria live in SOIL, and the blue-green algae are found in water, including the OCEANS. Plants absorb the nitrogen compounds and, in turn, pass the fixed nitrogen through the FOOD CHAIN, to animals, FUNGI, and other organisms. Some plants, especially many members of the pea family, or LEGUMES, have developed a form of MUTUALISM with nitrogen-fixing bacteria. The plants grow nodules on their roots that provide a home for the bacteria, and the bacteria provide the plants with fixed nitrogen. Farmers, especially those practicing ORGANIC FARMING, often plant a legume, such as clover, in order to build up nitrogen in a field that has been depleted by successive plantings of a crop species.

RETURNING NITROGEN TO THE ATMOSPHERE

Besides the nitrogen-fixing bacteria, there are also bacteria that perform the opposite task. They take organic compounds and use them when growing in conditions where there is no OXYGEN, and produce nitrite. In this way, nitrogen returns to the atmosphere, ready to travel the cycle again.

HUMAN IMPACTS ON THE NITROGEN CYCLE

The burning of fossil fuels and the concentration of LIVESTOCK in feed-lots create large amounts of organic nitrogen compounds, such as ammonia and NITROGEN OXIDES like nitrite and nitrate. Some researchers estimate that human activities produce about half as much organic nitrogen as does all other biological nitrogen fixing. These additional amounts of nitrogen compounds contribute to ACID RAIN, SMOG, and other forms of AIR POLLUTION. However, the nitrogen does not usually travel far, since many of the compounds are absorbed by living things. Near industrial areas, the amount of fixed nitrogen being added to the air and water can cause plants that were previously limited by low nitrogen levels to grow faster and can actually change the composition of some BIOLOGICAL COMMUNITIES by encouraging plant species that previously did not survive well in the area.

Nitrogen Dioxide

A NITROGEN compound composed of one nitrogen atom and two OXYGEN atoms (NO_2). Nitrogen dioxide is one of several nitrogen-containing chemicals known as NITROUS OXIDES. Nitrogen oxides make up approximately 11% of all primary air pollutants. Environmentalists are concerned about nitrogen dioxide as a pollutant because it is associated with several environmental problems.

Nitrogen dioxide is produced from the burning of FOSSIL FUELS, such as gasoline, oil, and COAL. The main sources of nitrogen dioxide released into the air are emissions from AUTOMOBILES, factories, and power plants. Nitrogen dioxide is one of the major components

◆ Scrubbers have been added to smokestacks of power plants to reduce nitrogen dioxide emissions.

of SMOG, a brownish haze that often forms over heavily populated and industrialized areas. Smog results from chemical reactions involving sunlight, nitrogen oxides, and HYDROCARBONS. In humans, smog can cause eye and lung irritation and possibly CANCER.

When nitrogen dioxide is released into the ATMOSPHERE, it can contribute to the formation of ACID RAIN. Acid rain, highly acidic rainfall, commonly forms in areas surrounding large cities and heavy industry. Acid rain and other forms of acid precipitation (snow, fog, or dew) can seriously harm the ENVIRONMENT.

Today, several laws are in effect to reduce nitrogen dioxide emissions. According to the CLEAN AIR ACT of the United States, smokestacks of power plants and factories must be equipped with SCRUBBERS. Scrubbers are devices that use water to remove nitrogen dioxide and other harmful chemicals from smoky emissions given off by burning fossil fuels. All new automobiles are equipped with CATALYTIC CONVERTERS. These devices reduce nitrogen oxide emissions by converting nitrogen oxides into less harmful substances. [*See also* AIR POLLUTION; HEALTH AND DISEASE; and WEATHERING.]

Nitrogen Fixing

▌The changing of NITROGEN in the air to nitrogen compounds such as nitrite or nitrate. Microorganisms, lightning, and meteorites all can fix nitrogen. However, microorganisms contribute 90% of fixed nitrogen. Nitrogen-fixing microorganisms include BACTERIA that live within nodules on the roots of PLANTS called LEGUMES, bacteria living in the SOIL, and certain ALGAE and LICHENS.

The most common nitrogen fixing system is the association of *Rhizobium* bacteria and legume (pod producing) plants. The bacteria enter the roots of the legumes. As the bacteria grow larger, they form swollen nodules on the roots. Within these nodules, the bacteria change nitrogen gas to nitrogen compounds that enrich the soil for plants to grow. [*See also* BIOGEOCHEMICAL CYCLE; CHEMICAL CYCLES; and NITROGEN CYCLE.]

Nitrogen Oxide

▌A compound, represented by the chemical symbol NO_x, formed when atmospheric nitrogen combines with OXYGEN in a process called *oxidation*. Lightning or cosmic-ray bombardment may trigger this process, along with NITROGEN FIXING, BACTERIA, and blue-green ALGAE. It also occurs in internal combustion engines, which produce 41% of nitrogen oxides, and industrial furnaces, which produce another 55%. The main nitrogen oxides are NITROGEN DIOXIDE and nitric oxide. Both cause health and

◆ Legumes, such as red clover, have nitrogen-fixing bacteria in their roots. The bacteria take nitrogen from the air in the soil and change it into a form usable by the plant.

environmental damage by forming ACID RAIN and photochemical SMOG and by stimulating OZONE formation in Earth's surface. However, living organisms need nitrogen oxides such as nitrite and nitrate to survive, and these compounds exist as part of the natural NITROGEN CYCLE. [*See also* AIR POLLUTION; OZONE POLLUTION; and PARTICULATES.]

Noise Pollution

▮A̲ny sound that interferes with hearing or has other harmful physiological or psychological effects on human health and welfare. Many sounds associated with human activities contribute to noise pollution. For example, heavy machinery, motor vehicles, aircraft, appliances, loud music, lawn and garden equipment, power tools, firearms, and even some toys produce sounds that can harm the human ear or otherwise be harmful to people. The harmful effects of sounds are related to their loudness or pitch.

HOW LOUD IS IT?

Sound is produced when molecules are set in motion. How "loud" a sound seems when it strikes the ear depends on the intensity, or the amount of energy, in the sound. The more intense a sound is, the greater the energy it has, and the louder it seems.

Intensity, or sound level, is measured in decibels (dB). The

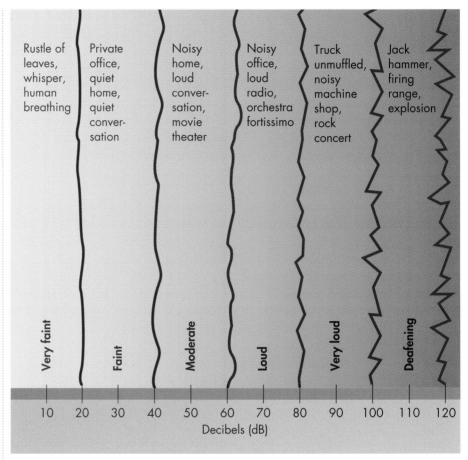

Very faint	Faint	Moderate	Loud	Very loud	Deafening
Rustle of leaves, whisper, human breathing	Private office, quiet home, quiet conversation	Noisy home, loud conversation, movie theater	Noisy office, loud radio, orchestra fortissimo	Truck unmuffled, noisy machine shop, rock concert	Jack hammer, firing range, explosion

10 20 30 40 50 60 70 80 90 100 110 120
Decibels (dB)

◆ The decibel scale is used to measure sound level.

lowest sound pressure that the human ear can detect has a sound level of 0 dB. Like the Richter scale for earthquakes, the decibel scale is logarithmic (increases in increments of powers of ten). Thus, a small increase in the number of decibels represents a large increase in intensity. For example, a sound measuring 1 dB produces ten times more energy than a sound measuring 0 dB. A sound measuring 2 dB produces 100 times more energy than a sound of 0 dB. A 3-dB sound produces 1,000 times as much energy as a 0 dB sound.

The ear does not perceive energy differences in sounds of different decibels. For example, in the same proportions, a sound measuring 3 dB seems to be only twice as loud as a sound measuring 2 dB, rather than ten times louder. A sound that measures 3 dB may seem only ten times as loud as a sound measuring 1 dB.

The noise produced by a motorcycle or a nearby subway can range from 75 to over 100 dB. In contrast, the average background noise in a typical home today is estimated to be between 40 and 50 dB. The music you hear at a rock concert or through headphones at a high setting may subject your ears to sound levels of 130 to 140 dB.

EFFECTS OF NOISE POLLUTION

The most obvious and measurable effect of noise pollution is damage to hearing. Exposure to high-intensity sounds for short periods of time can produce temporary hearing loss; longer exposure can result in permanent hearing loss. If you are exposed to noise levels of 75 to 80 dB for several hours, you might expect a temporary change in the range of sounds you can hear. Once the source of the noise is removed, however, your hearing will eventually return to normal. Short-term exposure to levels of between 130 and 140 dB can produce pain and a "ringing" sensation in the ears that may last for some time after the source is removed. Levels greater than 150 dB can cause physical damage inside the ear and permanent hearing loss. Loud noise may also affect other body systems. For example, other effects of continued exposure to loud sounds include irritability, headaches, fatigue, stress, and sleep disruption.

TAKING ACTION

The Noise Control Act of 1972 was enacted to promote an ENVIRONMENT free of harmful noise. It was the first federal attempt to control noise pollution in the United States. At one time, the ENVIRONMENTAL PROTECTION AGENCY (EPA) established an Office of Noise Abatement Control. Its duties included identifying the types of noises requiring control, setting standards, and educating people about noise pollution and safe noise levels. The office was disbanded in 1981, leaving noise regulation mainly in the hands of individuals and local agencies. In the workplace, the monitoring of noise levels is regulated by a federal agency called the Occupational Safety and Health Administration (OSHA).

As an individual, you can buy and use products that do not make noise. You can avoid noisy environments and wear protection when necessary. You can turn down the sound level on your radio, television, and stereo equipment. You can urge state and local governments to impose limits on noise levels. In addition, you can help to educate others about the potential dangers of noise pollution.

◆ A home is a typical nonpoint source. Many of the products people use at home can eventually contaminate water supplies.

Nonpoint Source

◗ A dispersed and often hard-to-identify source of POLLUTION of the air, water, or soil. Nonpoint pollution can come from anywhere. Homes, lawns, streets, farms, and gardens are typical nonpoint sources. When it rains, lawn fertilizers, household chemicals, animal droppings, and oil from cars all wash into storm sewer systems. From the sewer systems these pollutants can travel to and pollute OCEANS, rivers, and lakes.

According to studies by the ENVIRONMENTAL PROTECTION AGENCY (EPA), most freshwater bodies in the United States have already been affected by nonpoint pollution. Coastal areas are the most significantly affected. Pollution of coastal areas can result in contaminated

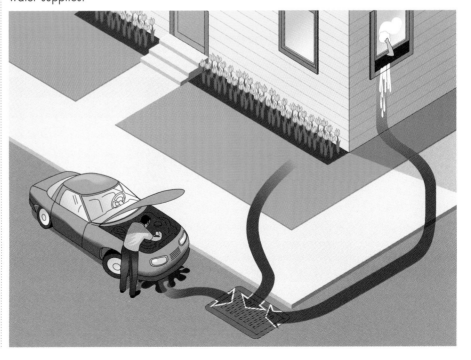

FISH and shellfish. It may also cause the temporary closing of beaches.

Because nonpoint pollution enters the ENVIRONMENT in more than one way, it is very difficult to control and regulate. Perhaps the most effective way to control such pollution is to educate people about the potential dangers of household chemicals and other products to the environment. [*See also* ALGAL BLOOM; CLEAN AIR ACT; CLEAN WATER ACT; ENVIRONMENTAL EDUCATION; EUTROPHICATION; HERBICIDE; INSECTICIDE; MARINE POLLUTION; PESTICIDE; RODENTICIDE; WETLANDS; and WETLANDS PROTECTION ACT.]

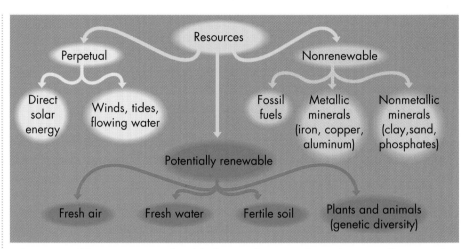

◆ Nonrenewable resources are fossil fuels, metallic minerals, and nonmetallic minerals. They are in limited supply.

Nonrenewable Resources

❏ NATURAL RESOURCES that are used by people faster than they are replaced by nature. By definition, nonrenewable resources are in limited supply on Earth. In contrast, RENEWABLE RESOURCES are those that are continually produced. Some renewable resources, such as sunlight, wind, and the energy of ocean waves, are so abundant they are considered available for an infinite amount of time.

When people talk about nonrenewable resources, they usually are referring to FOSSIL FUELS, such as COAL, PETROLEUM, and NATURAL GAS. Fossil fuels are the chief source of energy in the United States and most other nations. However, their supplies are dwindling rapidly. Fossil fuels are used mostly in operating AUTOMOBILES, power plants, and factories. Each of these practices makes use of fossil fuels at a faster rate than the millions of years it took for these FUELS to form. For example, coal is a fossil fuel that supplies about 28% of the world's energy needs. Coal is the most abundant fossil fuel on Earth; however, at current rates of demand, world coal supplies are expected to last only another 200 to 300 years. Supplies of natural gas and petroleum, or oil, are even more limited. Some experts estimate that Earth's supplies of these important fuels could be emptied within 50 years if current rates of use continue.

There are other problems associated with fossil fuels. The burning of fossil fuels for energy damages the ENVIRONMENT and contributes greatly to AIR POLLUTION. Because of these problems, scientists are developing ways to use the cleaner, renewable sources of available energy, such as WIND POWER, SOLAR ENERGY, and HYDROELECTRIC POWER. [*See also* ALTERNATIVE ENERGY SOURCES; ALUMINUM; ASBESTOS; BEST AVAILABLE CONTROL TECHNOLOGY (BACT); BIOMASS; COGENERATION; CONSERVATION; CONTAINER DEPOSIT LEGISLATION; ENERGY EFFICIENCY; GASOHOL; GEOTHERMAL ENERGY; OCEAN THERMAL ENERGY CONVERSION (OTEC); OIL DRILLING; OIL POLLUTION; OIL SHALE; PEAT; PETROCHEMICAL; PLASTIC; RECYCLING, REDUCING, REUSING; SOLAR HEATING; SUSTAINABLE DEVELOPMENT; SYNTHETIC FUEL; TELKES, MARIA; and UNITED NATIONS COMMISSION ON SUSTAINABLE DEVELOPMENT.]

Northern Spotted Owl

❏ Large, nocturnal, predatory BIRD that lives in the dense, OLD-GROWTH FORESTS of the Pacific Northwest of the United States, particularly in Washington and Oregon. The northern spotted owl is a subspecies of the spotted owl. In 1990, the U.S. FISH AND WILDLIFE SERVICE recognized that the HABITAT of the northern spotted owl was rapidly being lost. The HABITAT LOSS was a result largely of the cutting down of trees by the logging industry. Because of this habitat loss, the population of northern spotted owls was decreasing. In 1990, the population size had decreased so much that the U.S. Fish and Wildlife Service listed the northern spotted owl as a threatened SPECIES. Since that time, the northern spotted owl has been at the center of a heated controversy about how PUBLIC LANDS

of the United States should be managed and used.

Under the ENDANGERED SPECIES ACT of 1973, the habitats of ENDANGERED SPECIES—organisms that are likely to become extinct if critical factors in the ENVIRONMENT are not changed—must be protected. Similar action is taken to protect threatened species so they will not become endangered. Northern spotted owls became a species protected by the government. These birds nest only in trees of old-growth forests. An old-growth forest is one in which many of the trees in the forest have reached a stable configuration with a high canopy, an understory, and well-developed shrub and herb layers.

Such FORESTS represent a CLIMAX COMMUNITY.

In June 1990, when the United States listed the northern spotted owl as a threatened species, the owl became a symbol of how the economy and the environment can sometimes be at odds with each other. The owls became protected under federal law. Thus, logging was prohibited in the areas where the northern spotted owls lived. Loggers, sawmill workers, and residents of surrounding communities protested. They claimed that the recovery plan for the northern spotted owl would result in the loss of thousands of jobs and millions of dollars in logging revenues. They also argued that logging such forests and replanting them with young trees promotes the development of healthier forests. Critics of the northern spotted owl recovery plan felt that the livelihood of thousands of Americans was more important than saving a single species.

Environmentalists argue that the debate is not limited to only the northern spotted owl. They believe that it is also about protecting the remaining old-growth forests of the United States. These forests are rich in BIODIVERSITY. If logged, it may take hundreds of years for these complex ECOSYSTEMS to fully recover. Environmentalists also emphasize that these forests do not only provide a home for PLANTS and animals. They also help to reduce AIR POLLUTION and soil EROSION. Finally, they argue that automation and advances in logging technology have already caused a loss in logging jobs. Thus, changes in the logging

◆ The northern spotted owl was listed as a threatened species in 1990.

industry will continue for many years to come. To address such problems, environmentalists suggest that the logging industry make adjustments now while there is still some old-growth forest left.

The controversy over the northern spotted owl is not over. It will likely continue in the future. However, in 1995, the Supreme Court of the United States voted to uphold the recovery plan for the northern spotted owl. [*See also* CLEAR-CUTTING; CONIFEROUS FOREST; DEFORESTATION; ENVIRONMENTAL ETHICS; FORESTRY; FOREST SERVICE; KEYSTONE SPECIES; and MULTIPLE USE.]

No-Till Agriculture

▌▶Farming practice in which the seeds of the previous crop are planted alongside its remains. Tillage refers to the mechanical manipulation of SOIL by such methods as plowing, loosening, rolling, and packing. When soil is tilled, dead PLANTS are removed. In no-till agriculture, soils are not tilled at the end of the growing season to remove dead plants. Instead, a special machine cuts slits in the soil through the layers of dead plants. Seeds of the next crop are then planted alongside the old crops.

NO-TILL FARMING AND THE ENVIRONMENT

No-till farming has several important advantages over traditional

◆ No-till agriculture prevents erosion because soils are never left exposed to wind and rain. Instead, new crops are grown next to the dead crop plants.

farming practices in which soils are tilled. The main benefit is that soils are never left bare and exposed to wind and water EROSION. In traditional farming methods, erosion can be a constant problem. Plowing produces a loose layer of TOPSOIL that easily can be blown away by wind or washed away by rain. In no-till farming, the roots of the first crop hold soils in place while the new plants grow, thus preventing erosion. The American Soybean Association estimates that no-till farming can cut erosion by as much as 98%.

No-till farming can also improve

soil quality. As the old crops decay, organic matter is added to the soil, increasing soil fertility. In traditional farming, when the old plants are removed, soil fertility is reduced because the nutrients in the plants are not returned to the soil.

No-till farming can save time and money. No-till agriculture and other low-input farming methods, such as the recycling of animal manure and INTEGRATED PEST MANAGEMENT (IPM), seek to increase farm profits and protect the ENVIRONMENT by limiting the use of water, PESTICIDES, and chemical fertilizers. The use of no-till farming is expanding rapidly in the United States. According to the U.S. DEPARTMENT OF AGRICULTURE, no-till farming is used on more than 100 million acres (40 million hectares) of farmland in the United States. [*See also* AGRICULTURAL REVOLUTION; AGROECOLOGY; SOIL CONSERVATION; and SUSTAINABLE AGRICULTURE.]

Nuclear Energy

See NUCLEAR FISSION and NUCLEAR FUSION

Nuclear Fission

▎The process of splitting atoms to release energy. The process of nuclear fission was discovered in the 1930s by German physicists Otto Hahn and Fritz Strassman, working with Lise Meitner and Otto Frisch. The scientists achieved nuclear fission working with the radioactive element URANIUM. Several years later, nuclear fission was put to use to produce the first NUCLEAR WEAPONS. Today, this process forms the basis of energy production in NUCLEAR POWER plants.

Nuclear fission is best understood by studying the structure of an atom. Atoms are the smallest units of matter. The center of an

◆ In the 1950s, the Soviets studied atom smashers at the Institute Research Center in Debora, Russia.

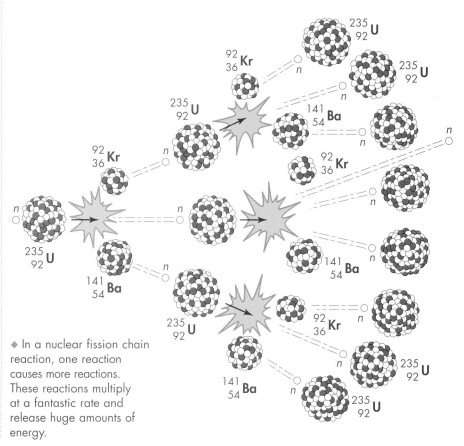

◆ In a nuclear fission chain reaction, one reaction causes more reactions. These reactions multiply at a fantastic rate and release huge amounts of energy.

atom, or *nucleus*, is made up of particles called *neutrons* and *protons*. In most chemical elements, these particles are held tightly together by a strong force. This energy, called *nuclear energy*, holds the nucleus of an atom together.

Some chemical elements, such as uranium, are radioactive. A radioactive atom is unstable and releases neutrons and protons from its nucleus. When this happens, the nuclear energy that usually keeps these particles together is also released. In its simplest sense, this release of energy is nuclear power.

To bring about nuclear fission, scientists bombard the nuclei of unstable radioactive elements with neutrons. When a neutron strikes the nucleus of a radioactive atom, it causes the nucleus to break apart. This splitting, or fission, releases huge amounts of energy, as well as the protons and neutrons of the nucleus. If some of the neutrons are allowed to bombard other radioactive atoms, a chain reaction will occur, in which each reaction causes more reactions.

In a chain reaction, the energy released grows at a fantastic rate. Suppose, for example, a neutron strikes a radioactive nucleus causing the release of two neutrons. If these two neutrons are allowed to strike other radioactive atoms, there would be about 1,000 fission reactions after only one one-millionth of a second. (After 2 milliseconds, 1 million fission reactions would occur. After 3 milliseconds, there would be a billion reactions. Thus, in a fraction of a second, enormous amounts of energy would be released. This type of uncontrolled chain reaction forms the basis of the atomic bomb.)

In a nuclear power plant, such violent explosive chain reactions are not desirable. Instead, chain reactions are controlled inside a nuclear reactor. In the reactor, devices called *control rods* absorb some of the neutrons released during the chain reaction. In this way, a chain reaction produces only as much energy as needed. Nuclear power plants harness the energy produced during the fission reactions and use it to generate ELECTRICITY. Similarly, controlled chain reactions are used to generate energy and additional radioactive FUEL in a BREEDER REACTOR. [*See also* PLUTONIUM; RADIATION; RADIOACTIVE WASTE; and RADIOACTIVITY.]

Nuclear Fusion

▮ The joining, or fusing, of the nuclei of two **atoms** to produce energy as well as the nucleus of a new element. In fusion, the nuclei of two hydrogen **isotopes** fuse to form the nucleus of one helium atom. Nuclear fusion occurs in nature, but not on Earth. This type of reaction produces the energy in stars, including our sun. Fusion

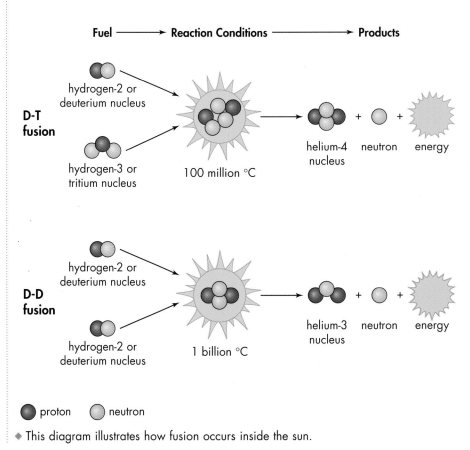

Fuel ⟶ Reaction Conditions ⟶ Products

D-T fusion
hydrogen-2 or deuterium nucleus
hydrogen-3 or tritium nucleus
100 million °C
helium-4 nucleus   +   neutron   +   energy

D-D fusion
hydrogen-2 or deuterium nucleus
hydrogen-2 or deuterium nucleus
1 billion °C
helium-3 nucleus   +   neutron   +   energy

● proton   ○ neutron

◆ This diagram illustrates how fusion occurs inside the sun.

occurs constantly inside the sun to produce tremendous amounts of energy that is given off as light and heat. Many scientists believe that nuclear fusion may someday supply all the world's energy needs.

Nuclear fusion is similar to NUCLEAR FISSION, the splitting of atoms, in that both processes release enormous amounts of energy. However, nuclear fusion produces

more energy than nuclear fission in a less costly and safer way. In addition, the fusion process does not produce nuclear waste. However, nuclear fusion is more difficult to produce.

Scientists have had difficulty doing experiments with nuclear fusion ever since they began studying it in the 1930s. Problems occur because nuclear fusion requires temperatures of more than 200 million ° F (100 million ° C). Thus, one of the main problems in carrying out nuclear fusion has been to find a container able to withstand such extreme temperatures.

In the 1950s, American researchers produced the hydrogen bomb. This bomb works under the principle of nuclear fusion. To obtain the high temperatures needed, the scientists used the heat produced from the explosion of an atomic bomb. Since that time, scientists have been searching for more practical ways to produce the heat needed to fuse atoms.

Large-scale nuclear fusion experiments have been conducted in various countries, including the United States and the former Soviet Union. Researchers also continue to study the fusion reactions that occur in the sun. More recently, scientists have experimented with a process called *cold fusion*. This process merges atoms in water at room temperature. Cold fusion would provide an inexpensive, safe, and plentiful source of energy. However, many scientists feel that cold fusion will not be achieved for many years. [*See also* ALTERNATIVE ENERGY SOURCES; NUCLEAR WEAPONS; RADIATION; and RADIOACTIVITY.]

◆ Scientists are experimenting with nuclear fusion as a source of energy.

Nuclear Power

Energy produced by the splitting of the nucleus of an atom into smaller nuclei (NUCLEAR FISSION) or through the joining of two small nuclei to make a larger one (NUCLEAR FUSION). As atomic nuclei are changed, they release enormous amounts of heat energy. This energy can be harnessed and used to produce steam. The steam can be used to power machinery, generate ELECTRICITY, or propel a vehicle.

Nuclear power plants create electricity by splitting the nuclei of atoms of heavy elements such as URANIUM in a device called a *nuclear reactor*. Heat produced by this process is used to change liquid water into steam. The steam can then be used to rotate the turbine of an electric generator. Once steam is used, it can be condensed back into liquid water. Thus, water can be recycled and used again. Most nuclear power plants use giant cooling towers to allow water from condenser pumps to cool down.

NUCLEAR POWER PLANTS — GOOD NEWS AND BAD

Nuclear power plants produce approximately 16% of the world's electricity. About 19% of the energy produced in the United States results from nuclear power. An advantage of using nuclear power plants to generate electricity is that they use less FUEL than do plants that burn FOSSIL FUELS to produce equal amounts of electricity. In addition, nuclear power plants create less AIR POLLUTION than power plants that make use of fossil fuels.

The use of nuclear power has some disadvantages. One is that the radioactive material used in nuclear plants produces RADIOACTIVE WASTE. These wastes create a disposal problem because nuclear wastes can remain radioactive for thousands of years or longer. In addition, nuclear power plants cost more to build than do power plants

◆ Although nuclear power plants produce radioactive wastes, they remain a major source of electricity.

that use fossil fuels. The difference in expense is due to special safety systems that are needed to prevent accidental discharge of excess RADIATION into the ATMOSPHERE and to quickly contain radiation should there be an accident such as that which engineers call the China syndrome. In this situation, the reactor core becomes very hot and melts the structure that contains it, resulting in the release of radioactive material into the ENVIRONMENT.

FROM THREAT TO REALITY

Every nuclear plant releases some radiation into the environment, but generally too little or not enough to be harmful. However, in the United States, there were minor accidents before 1979 in which more radiation than normal was emitted. Then on March 28, 1979, a major accident took place at the THREE MILE ISLAND nuclear power plant near Harrisburg, Pennsylvania. The threat posed by this accident caused many people to question the safety of nuclear power. In 1986, the world's worst nuclear power plant disaster occurred at CHERNOBYL in the former Soviet Union. As a result of this accident, many people died immediately. Others living near the area have been left with illnesses that will affect them for the rest of their lives. To prevent future accidents, engineers continue to design improvements for nuclear reactors, such as using enclosures of graphite and layers of ceramics and CARBON rather than the metal enclosures presently used. However, concern over the possibility of such accidents has led the United States to abandon construction of new nuclear power plants. [*See also* ATOMIC ENERGY COMMISSION (AEC); BREEDER REACTOR; CANCER; CARCINOGEN; PLUTONIUM; THERMAL WATER POLLUTION; and URANIUM.]

◆ Nuclear energy is used to generate electricity as well as to power some naval vessels, including submarines.

Nuclear Weapons

▌▌Defensive devices, such as bombs, missiles, or torpedoes, that produce explosive energy from NUCLEAR FISSION or NUCLEAR FUSION. In nuclear fission, atoms are split. The two most common elements used in fission reactions are the radioactive elements URANIUM and PLUTONIUM.

The nuclear bombs dropped on Japan during World War II were fission weapons. Nuclear fission is also used to produce electrical energy in NUCLEAR POWER plants.

Fusion weapons produce energy from nuclear fusion. In this process, atoms are fused together, rather than split apart. The hydrogen bombs developed in the 1950s were fusion weapons. Nuclear fusion also produces the tremendous amounts of light and heat given off by stars, such as our sun.

◆ Nuclear weapons are stockpiled by the major powers, including the United States and Russia.

Nuclear weapons cause great destruction to people and the ENVIRONMENT. Explosions from nuclear weapons create extremely high temperatures, which can burn objects instantly and reduce buildings to rubble in seconds. Exposure to high levels of RADIATION from a nuclear blast can cause CANCER and other illnesses in people. In addition, some scientists fear that a large-scale nuclear war could lead to a widespread CLIMATE CHANGE known as NUCLEAR WINTER. Recognizing the destructiveness of nuclear weapons, many of the world's nations have been trying to reach agreements about limiting how many of them can be built. [*See also* NUCLEAR POWER; RADIATION EXPOSURE; RADIOACTIVE FALLOUT; and RADIOACTIVITY.]

Nuclear Winter

▎Term used to describe the possible environmental and climatic effects that would result from large-scale nuclear war. The term "nuclear winter" was first coined by American physicist Richard P. Turco in an article that appeared in the journal *Science* in 1983. In this article, Turco argued that the billions of tons of dust, soot, smoke, and ash that would rise from burning cities following a nuclear war could block at least 80% of the sunlight that reaches the Northern Hemisphere. As a result, severe worldwide CLIMATE CHANGES could occur. Such changes might include prolonged periods of darkness, below-freezing temperatures, violent windstorms, and RADIOACTIVE FALLOUT. According to this view, the combination of cold temperatures, dryness, and lack of sunlight would slow agricultural production and destroy ECOSYSTEMS. These changes would place the majority of the world's populations at risk of starvation. [*See also* CANCER; CLIMATE; NUCLEAR WEAPONS; RADIATION; RADIATION EXPOSURE; and RADIOACTIVITY.]

Nutrient Cycle

See BIOGEOCHEMICAL CYCLE; NITROGEN CYCLE; and OXYGEN CYCLE

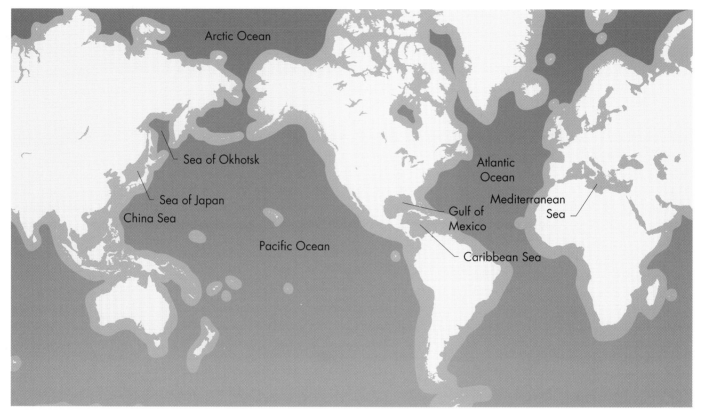

Arctic Ocean

Sea of Okhotsk

Sea of Japan

China Sea

Pacific Ocean

Atlantic
Ocean

Mediterranean
Sea

Gulf of
Mexico

Caribbean Sea

◆ The region of ocean within a few miles of shore is called the neritic zone, which is rich in marine life. The open ocean contains relatively few organisms.

Ocean

▶A large, continuous body of salt water contained in deep basins on Earth's surface. Oceans cover about 140 million square miles (360 million square kilometers), or nearly three quarters, of Earth's total surface area.

The world's oceans are commonly divided into several major oceans and smaller seas. The three principal oceans are the Atlantic Ocean, the Pacific Ocean, and the Indian Ocean. These vast oceans also connect with what is sometimes called the Southern Ocean, the body of water that surrounds the Antarctic continent. Most of the smaller seas occur in the Northern Hemisphere and are enclosed by land or surrounded by island systems. These lesser oceans include the Arctic Ocean, the Mediterranean, the Caribbean, and the Bering Sea, among others.

EXPLORING OCEANS: THE ROLE OF OCEANOGRAPHY

Scientists who study oceans are known as *oceanographers*. The science of oceanography is important for describing and mapping the features of the ocean floor and understanding the physical features and chemical properties of oceans. Oceanographers also study how the composition and the temperature of ocean water affect marine organisms.

THE STRUCTURE OF THE SEAFLOOR

Scientists have a good picture of what Earth's surface looks like beneath the oceans. Using special instruments, such as sonar and diving vessels, scientists can map every bump and groove on the ocean floor. While mapping the ocean, scientists have made some startling discoveries about this unseen portion of Earth.

Generally, seafloor features are similar in all oceans. At the edge of the ocean are vast continental shelves and slopes that extend out from the shoreline. The bottom of oceans is the seafloor. Most of the seafloor is flat, forming extensive areas called *abyssal plains*. However, in some areas, this flat bottom is disrupted by huge underwater mountain chains, deep valleys, and underwater volcanoes.

Away from the continents, oceans can be very deep. Average depth of the world's oceans is about 2.3 miles (3.7 kilometers). However, scientists have discovered much deeper areas. The deepest parts of the ocean occur in huge crevices called *ocean trenches*. These long, deep trenches are caused by PLATE TECTONICS, where one section of the sea floor is being pushed under another section. The deepest ocean trench is the Marianas Trench in the Pacific Ocean. At 36,198 feet (11,033 meters) below the surface, it is the deepest place on Earth—almost seven times deeper than the Grand Canyon!

OCEAN WATER: COMPOSITION AND TEMPERATURE

Oceanographers have learned that ocean water contains many dissolved chemicals. These include sodium, chlorine, silica, and calcium. The two most abundant chemicals are sodium and chlorine in the form of sodium chloride, a substance known as table salt. Sodium is dissolved in river waters that flow into oceans. Chlorine is added by underwater volcanoes. Together, these two chemicals combine to form a salt called sodium chloride or *halite*.

Scientists have determined that ocean water contains about 3.5% salt. In areas with large amounts of rainfall, rivers, or melting ice, the SALINITY is lower where ocean water mixes with fresh water. However, in some areas, such as the Persian Gulf and the Red Sea, salt content can reach 4.2%. This higher salt content is partially due to evaporation of water and RUNOFF from mineral-rich land. The salinity of ocean water in most areas has stayed about the same over the past 600 million years. This suggests that the composition of ocean water is in balance.

The temperatures of the SURFACE WATERS of Earth's oceans and seas varies slightly according to location. The sun is the main supplier of heat for Earth's oceans. For instance, in polar areas that do not receive much sunlight, surface waters can be as low as 29° F (−1.7° C). In areas that receive more direct sunlight, such as those around the equator, ocean temperatures may

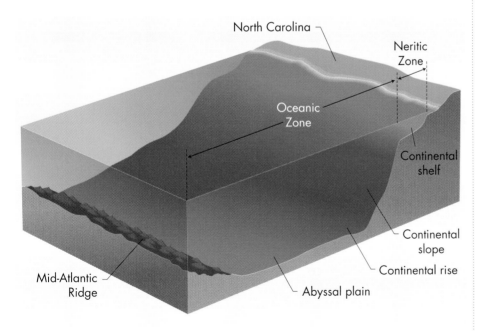

◆ At the edge of the oceans are continental shelves that gently tilt down from the land. Farther out, the shelves become continental slopes that drop sharply down to the ocean depths.

rise to as much as 86° F (30° C). Generally, temperatures decrease with increasing depth, except in polar waters, which show the reverse pattern.

OCEAN ECOSYSTEMS

More animals and PLANTS live in the world's oceans than anywhere else. However, most life is concentrated near the surface along the edges of continents. Marine ECOSYSTEMS are divided into two distinct groups, depending upon their distance from the shore: the coastal region, or *neritic zone,* and the open ocean, or *oceanic zone.*

Neritic Zone

The neritic zone is the region of the ocean that exists within a few miles of shore above the continental shelf. Here, the shallow, warmer waters and the large amounts of nutrients washed in from the land permit the growth of large seaweed anchored to the bottom, and of microscopic ALGAE, or PHYTOPLANKTON, floating near the surface.

In turn, these tiny PRODUCERS attract a great variety of animal SPECIES. These include many types of FISH and BIRDS; MAMMALS, such as sea otters, DOLPHINS, and manatees; and a great variety of INVERTEBRATES, such as jellyfish, corals, crabs, shrimp, and sea stars.

Oceanic Zone

Life is much less diverse in the open ocean, or oceanic zone. This zone exists well beyond the continental shelf. Here, life is primarily concentrated close to the surface where sunlight can penetrate the water. Since light can only penetrate ocean water about 590 feet (180 meters), the primary producers in the oceanic zone ecosystem are floating phytoplankton. The abundance of phytoplankton attracts a variety of fish and marine mammals. These animals attract larger PREDATORS, such as sharks and killer WHALES.

Few organisms live deep below the surface in the oceanic zone. Except in **hydrothermal vent** regions, most deep water organisms are scavengers and DECOMPOSERS that depend on the constant downward flow of dead organisms and nutrients. Many types of deep-sea predators, such as the angler fish, have been discovered. Scientists are also finding new species living on the floor of the ocean, even at depths of 5 miles (8 kilometers).

OCEANS AND HUMAN SOCIETY

In addition to serving as HABITAT for more than 250,000 different species of organisms, oceans play other important roles for life on Earth. For instance, the large amounts of algae

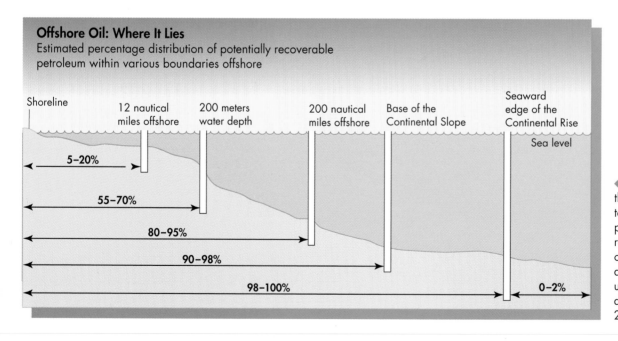

Offshore Oil: Where It Lies
Estimated percentage distribution of potentially recoverable petroleum within various boundaries offshore

Shoreline

12 nautical miles offshore

200 meters water depth

200 nautical miles offshore

Base of the Continental Slope

Seaward edge of the Continental Rise

Sea level

5–20%

55–70%

80–95%

90–98%

98–100%

0–2%

◆ It is estimated that from 50% to 70% of the potentially recoverable offshore oil lies at the sea bottom under water no deeper than 200 meters.

hydrothermal vent an opening on the sea floor from which hot solutions rich in minerals flow.

land-based coming from coasts, ground, or soil.

living near the surface of the ocean supply much of the OXYGEN used by both land and sea organisms. Oceans also help regulate WEATHER and CLIMATE by contributing to the global WATER CYCLE, and by distributing stored heat through OCEAN CURRENTS.

The oceans and the organisms they contain are also very important to humans. Humans rely on oceans for recreation and for transporting goods throughout the world. Oceans also supply people with food. In the United States, seafood is a $10 billion industry. Fish products and other marine organisms are also used to make soaps, medicines, cosmetics, animal feeds, fertilizers, and other commercial products.

Threats to Oceans

Although oceans are vast and contain many resources, these delicately balanced ecosystems are becoming polluted and damaged due to human activity. Whales, dolphins, fish, crabs, shrimp, clams and other marine species are being caught and harvested in record numbers. Accidental discharge of oil from tankers and the illegal dumping of GARBAGE and other wastes also damage the ocean.

Most sources of ocean POLLUTION are **land-based**. Pollution reaches the oceans when factories and towns discharge SEWAGE and industrial wastes into rivers, streams, and harbors. A significant amount of ocean pollution also comes from NONPOINT SOURCES. Nonpoint pollution is indirect pollution that comes from many sources. Water runoff from farms, cities, and construction sites often contains PESTICIDES, fertilizers, oils, paints, and other chemicals that collect along shores and poison organisms.

Saving Ocean Ecosystems

In the United States, a number of laws protecting oceans have been passed as public awareness of ocean pollution has grown. The 1977 CLEAN WATER ACT and the Ocean Dumping Ban Act passed in 1988 are important laws designed to keep ocean waters clean. The 1972 MARINE MAMMAL PROTECTION ACT protects dolphins, whales, sea otters, and other marine mammals that live in the waters of the United States.

On an international level, the International Convention for the Prevention of Pollution from Ships, known as MARPOL, was developed in 1973 to help control ocean pollution. The LAW OF THE SEA Convention, developed in 1982, is another important international agreement to control ocean pollution. Today, the LOSC has been signed by more than 160 nations who agree that deep-sea resources, such as minerals, represent "the common heritage" of humankind. [*See also* ESTUARY; FISHING, COMMERCIAL; FISHING, RECREATIONAL; INTERNATIONAL CONVENTION FOR THE REGULATION OF WHALING (ICRW); INTERTIDAL ZONE; MARINE POLLUTION; OCEAN DUMPING; OIL POLLUTION; PLATE TECTONICS; SALT MARSH; and WATER POLLUTION.]

Ocean Current

The regular movement of water along certain paths within the OCEAN. Ocean water does not stay in one place. Large masses of water constantly flow like huge rivers from one place to another within the ocean. Some currents are at the surface, while others flow far below the surface. At the border between two currents, someone in a boat or submarine will likely observe a change in the temperature, color, or direction of the water. A change in the organisms that live in the water may also be seen.

CAUSES OF OCEAN CURRENTS

Tides, severe storms, and even undersea earthquakes can all create currents that move water back and forth for short distances. The most important cause of currents are tides. Tides are caused by the pull of the moon's gravity on Earth's oceans. In coastal ENVIRONMENTS, many organisms depend on food that is carried along with the moving tides.

Some currents are caused by differences in density of ocean water. Water that is very cold or very salty, for example, is more dense than warmer water or water with low SALINITY. Thus, a certain volume of very cold or very salty water has more mass than an equal volume of warmer, fresher water. The denser water sinks and may flow along the ocean bottom. Less dense water tends to spread out over the surface of the ocean. These movements of more dense and less dense water create currents where cold and warm water meet or where fresh water meets salty water.

Other currents are caused by the dragging of wind along the surface of the water. In some parts of the world, the wind blows almost constantly in one direction. Just north and south of the equator, for example, there are fairly continuous winds called *trade winds*. The winds got their names from the fact that trading ships once depended on them for sailing across the seas. When wind blows over the ocean, it slowly pulls the surface along in the same direction. This creates surface currents that extend over long distances. Once currents are in motion, their speed and direction are also influenced a great deal by the **Coriolis effect**. This effect helps explain why ocean currents

tend to go around basins in great circles, or *gyres*.

EFFECTS OF OCEAN CURRENTS

Ocean currents are enormous. The Gulf Stream, one of the most famous North Atlantic currents, extends from the Gulf of Mexico to northern Europe. It carries about 100 times as much water as all the world's rivers put together. Another current (the Antarctic Circumpolar Current) goes all the way around the continent of ANTARCTICA. Its flow is about twice as great as the Gulf Stream.

These powerful streams within the ocean affect life all over the world. Some of the streams absorb a great deal of heat under the tropical sun. This heat is carried in the water for thousands of miles to the north or south. Without the warmth brought to the north by tropical ocean currents, countries such as England and Ireland would be much colder than they are. Currents also carry nutrients. In some parts of the ocean, cold currents come up from the bottom to the surface, carrying MINERALS that are much needed for the growth of PHYTOPLANKTON. The phytoplankton, in turn, support marine FOOD WEBS. Without these **upwellings**, the ocean would support less life. Currents also carry marine organisms. This allows microscopic organisms and sessile plants and animals to move long distances and settle in new HABITATS.

Pollutants are often carried in ocean currents. As people have learned about ocean currents, it has

become clear that wastes dumped in one part of the ocean can turn up again somewhere else. For

Coriolis effect a change in the direction of travel of objects moving on Earth (or any spinning sphere or disk), which arises from the fact that not all parts of Earth are spinning at the same rate.

To picture the Coriolis effect, imagine two objects on the spinning Earth, one near the Equator and one near the North or South Pole. The object near the North or South Pole is carried in a much smaller circle each day. The object at the Equator must therefore be moving faster than the object to the north or south. Now imagine what happens to water if it is first on the Equator, being carried around quickly with Earth's rotation from west to east, and then, heated by the sun, it spreads north or south away from the Equator. The water is moving into a region of Earth that spins from west to east more slowly than the equator does. However, the water does not slow down from its "Equator speed" quite enough to match the speed of the land beneath it. The water keeps going east a little faster than the ocean bottom does. This is why ocean currents flow away from the Equator to the northeast and southeast.

upwellings the upward movement to the surface of lower nutrient-rich waters.

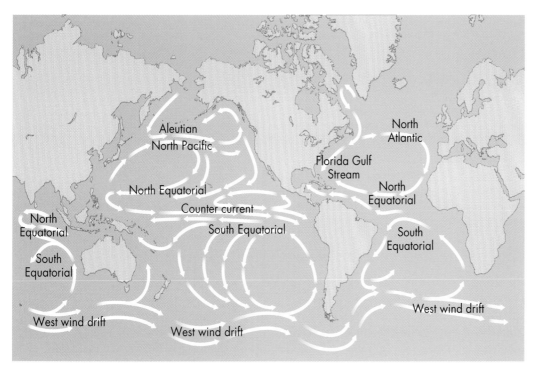

◆ Major ocean currents of the world. Many ocean basins have a circular pattern, or gyre, of ocean currents, with a strong flow on the west side of the basin and a more sluggish flow on the eastern side.

Aleutian
North Pacific

North Atlantic

Florida Gulf Stream

North Equatorial

North Equatorial

Counter current

North Equatorial

South Equatorial

South Equatorial

South Equatorial

West wind drift

West wind drift

West wind drift

example, wastes dumped in fairly shallow water just off the coast are likely to be moved by ocean currents. This tendency has caused concern about the long-term effect of OCEAN DUMPING on the marine environment. [*See also* MARINE POLLUTION and SEABED DISPOSAL.]

no problems or only local problems. However, today wastes are produced on a much greater scale, largely due to increases in human population sizes. In addition, to-

day's wastes include a larger variety of pollutants than did wastes of the past. Ocean dumping therefore presents greater problems than it did in the past. Problems associated

Ocean Dumping

❚ Direct disposal of SOLID WASTE and other pollutants into the ocean from ships, barges, or pipelines. For centuries, people have released SEWAGE and other wastes from cities and towns along seacoasts into the OCEAN. Usually, this practice caused

◆ Liquid waste containing phosphate and other chemicals may build up to high concentrations and become toxic.

with ocean dumping include dangers to the marine ENVIRONMENT and human health. These dangers have led to the passage of laws to limit dumping.

TYPES OF OCEAN DUMPING

Many kinds of ocean dumping were common in the United States until a short time ago. *Dredge soil*, material dug out of the bottoms of harbors and shipping channels, was usually dumped a few miles offshore. Other wastes that have been dumped into the ocean over the years include: sewage, SLUDGE, wastes from FISH canning companies, acids, PETROCHEMICAL wastes, wastes from steel and paper industries, PESTICIDES, ships' GARBAGE, outdated medicines, and rubble from torn-down buildings. Ashes have been dumped from incinerator ships, which burned wastes at sea. In addition, the U.S. government has dumped old military explosives, batteries that once powered navigational buoys, ammunition, and chemical weapons at sea. Sometimes, such wastes were dumped in old ships that were simply sunk at the dumpsite. Even RADIOACTIVE WASTES have been dumped deep in the sea.

EFFECTS OF OCEAN DUMPING

Some kinds of dumped waste have been used to provide new HABITATS for marine organisms. Many ALGAE, shellfish, fish, and other organisms live on natural reefs, which are underwater platforms of stone or coral. Blocks of cement, old car

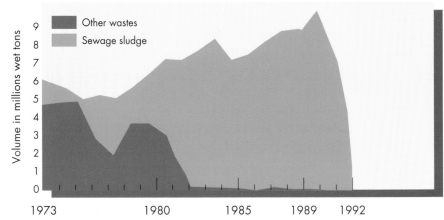

◆ The amount of sewage dumped offshore decreased dramatically in the early 1990s, while the amount of other wastes decreased ten years earlier.

◆ Ocean dumping in shallow coastal water affects marine life that people use for food.

bodies, or old tires can be made into artificial reefs, which can also support diverse populations of PLANTS and animals. Dredge soil that is not too heavily polluted can be used to build islands for WILDLIFE.

Most dumped pollutants, however, are dangerous to marine life. It was once believed that such pollutants could not do much harm in a large ocean. People believed that the "solution to pollution is dilution." Dumped toxins, for example, were supposed to be mixed with such large amounts of seawater over such a wide area that they could not poison marine life in any one spot.

Contrary to what people once thought, ocean dumping has turned out to be a problem. Dredge soil, sewage, and sludge that build up too quickly can bury the animals that live on the ocean bottom. Some sewage contains oil and other chemicals or BACTERIA and VIRUSES. This is not only bad for marine organisms, but it can bring toxins and diseases to people who eat oysters, clams, crabs, lobsters, and fish from the polluted ocean bottom. Pesticides and other chemicals in seawater can build up within the bodies of fish or marine mammals through the process of BIOACCUMULATION. Hospital wastes that are dumped illegally can wash up on beaches. This can injure people and spread disease.

LIMITS ON DUMPING

Many countries have decided to pass laws that limit ocean dumping. The Federal Water Pollution Control Act Amendments of 1972, the MARINE PROTECTION, RESEARCH, AND SANCTUARIES ACT of 1972, the Medical Waste Tracking Act of 1988, and the Ocean Dumping Ban Act of 1988 have stopped most ocean dumping by industries in the United States. However, ocean dumping is still a serious problem in many other parts of the world, especially in coastal areas. To help solve the problem, a number of countries have agreed to the rules of the International London Convention of 1972 (the Convention on the Prevention of Marine Pollution by Dumping of Wastes and Other Matter). The United Nations LAW OF THE SEA Convention also contains rules to limit ocean dumping worldwide. [*See also* MARINE POLLUTION; OCEANS; and WATER POLLUTION.]

Ocean Thermal Energy Conversion (OTEC)

▶A system producing ELECTRICITY by making use of temperature differences in ocean water. The limited availability of FOSSIL FUELS and their tendency to produce large quantities of AIR POLLUTION have led scientists to seek out alternative methods for generating electricity. One energy source being experimented with is ocean thermal energy conversion (OTEC). This process uses the temperature differ-

ences between warm SURFACE WATERS of the ocean and cold temperatures of waters 2,000 to 3,000 feet (about 600 to 900 meters) below the surface to generate electricity.

Surface waters of the ocean are heated as a result of SOLAR ENERGY. However, the sun's rays are unable to penetrate deep into ocean waters. As a result, temperatures there are much cooler than waters at the surface. In an ocean thermal energy conversion plant, warm surface waters come into contact with a closed tube that holds ammonia. The warmth from the water heats the ammonia in the tube, changing it to a gas. As the ammonia gas flows through the tubing, it rotates a **turbine** that is connected to a

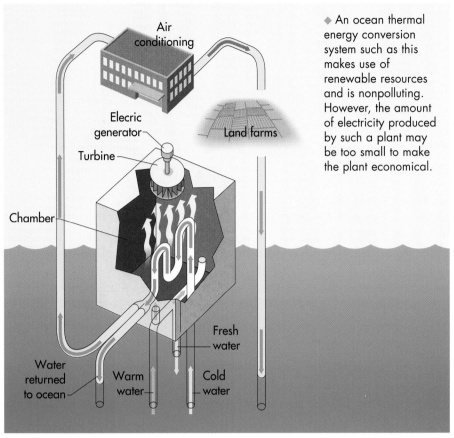

◆ An ocean thermal energy conversion system such as this makes use of renewable resources and is nonpolluting. However, the amount of electricity produced by such a plant may be too small to make the plant economical.

believe changing the temperature of water near OTEC plants may result in changes in WEATHER conditions of the area. [*See also* ALTERNATIVE ENERGY SOURCES and THERMAL WATER POLLUTION.]

Odum, Eugene Pleasants (1913–)

generator. In the generator, the mechanical energy of the moving turbine is changed into electricity.

After the ammonia gas moves past the turbine, it is cooled by water that is pumped up to the system from deep within the ocean. The cooling action of the water changes the ammonia back to a liquid. Because the ammonia is contained in a closed tube, it flows back to the area containing the warm surface waters, where it is once again changed to a gas, beginning the process again.

The use of OTEC would result in a continual source of electricity that makes use of RENEWABLE RESOURCES. However, thus far the system has not been shown to be practical. One major problem is that very large amounts of water are needed to produce only a very small amount of electricity. This gain in electricity is made even smaller because some of the electricity is used to pump cold ocean waters up near the surface for use by the system.

Scientists are also concerned that OTEC may pose some threat to the ENVIRONMENT. As cool ocean water passes over the tube of the system, it is released back to the ocean very near the surface. This cool water lowers the temperature of surrounding waters. Organisms living in the area may be unable to adapt to such changes in temperature. In addition, some scientists

❚ An ecologist widely recognized as one of the pioneers of modern ECOLOGY. Eugene Pleasants Odum was a member of an unusually talented family. His father was eminent sociologist Howard Washington Odum, and his brother, Howard T. Odum, is equally well known for his application of energy analysis and simulation modeling to ecological science.

E. P. Odum wrote a widely used textbook on ecology, *Principles of Ecology*. After World War II, he played a pivotal role in establishing the Institute of Ecology at the University of Georgia and the Savannah River Ecology Laboratory, which is located at the site of the Savannah River plant—one of three sites instrumental in producing atomic bombs. Odum wanted scientists to take advantage of the 300-square-mile (777-square-kilometer) Savannah River plant site, which was to be preserved for decades.

Odum has won numerous awards for his leadership in the field of ecology and for advancing the holistic systems approach to the study of ECOSYSTEMS. He pioneered the approach of studying systems for the way they process energy and materials, especially nutrients. Odum won the Tyler Prize and, with H. T. Odum, the Crafoord Prize given annually for ouastanding work in ecology. He is a member of the National Academy of Sciences and is respected as the person who motivated scientists to consider ecology one of the system sciences.

Office of Surface Mining, Reclamation, and Enforcement

▌A federal agency created in the DEPARTMENT OF THE INTERIOR by the Surface Mining Control and Recovery Act (SMCRA) of 1977. This act required mining companies to reclaim or restore landscapes that had been laid bare by strip-mining

operations. SMCRA established federal standards for these reclamation efforts and established the Office of Surface Mining to publish regulations. Under the provisions of SMCRA, the states were permitted to create their own laws governing reclamation efforts, thereby ensuring that each state would be able to deal with possibly unique local conditions and yet meet the federal standards. Thus, while companies would be required by federal law to reclaim mining sites on PUBLIC LAND, reclamation of sites on private lands could also be required by law in some states.

A further stipulation of SMCRA was that the Office of Surface Mining should be an independent federal regulatory agency. This stipulation was repealed, however, and the office's authority thereby weakened during James Watt's controversial tenure as Secretary of the Interior under President Ronald Reagan. The office's loss of funding and technical support corresponded with the granting of broad regulatory responsibility to the states.

Oil

See PETROLEUM

Oil Drilling

▌Method for recovering oil from underground deposits. In several

parts of the world, oil is present in vast underground pockets called *reservoirs.* Reservoirs may be located thousands of feet underground or closer to the surface. Most often, oil reservoirs exist in highly porous rocks, such as sandstone or limestone, that are sandwiched between layers of non-porous rocks.

To recover oil from an underground deposit, a deep hole must be drilled into the oil reservoir. A long recovery pipe is then driven into the earth. Usually, the oil flows out of the ground under its own pressure. This natural, upward flow of oil is caused by several different

◆ Oil mules are used to drill for oil at a California beach.

As land reserves decline, oil will be drilled at offshore platforms such as these used in the Gulf of Mexico.

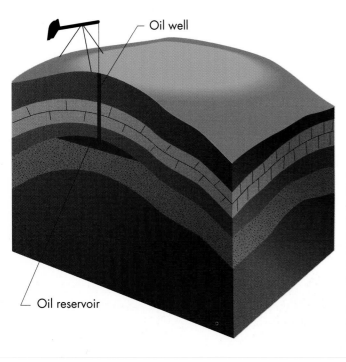

◆ Oil reservoirs may be located thousands of feet underground in porous rocks that are sandwiched between layers of nonporous rocks. Recovery pipes are inserted into holes drilled into the reservoir. Oil usually gushes up under its own pressure.

factors. In some cases, water pressure within the porous rock pushes oil upward and into the recovery pipe. In others, pressure caused by NATURAL GAS reservoirs forces oil through the recovery pipe. The recovery of oil that uses the natural flowing power of oil is known as *primary recovery*. Additional oil may be recovered from a reservoir by pumping, a method called *secondary recovery*.

Even after primary and secondary recovery methods have been used, some oil may still remain in a reservoir. In these instances, *tertiary recovery* techniques are used to remove the oil. These techniques

◆ Unrefined, or crude, oil is thick and does not pour at room temperature.

Where does all this oil come from? Nearly half is from the illegal and accidental discharge of oil from tankers, ships, and recreational boats. Also, offshore oil rigs sometimes leak oil during drilling. The rest of the oil comes from the land—from coastal oil refineries, factories, natural seepage from underground oil deposits, and the illegal dumping of motor oil by individuals.

Oil pollution can have devastating effects on marine WILDLIFE. Small animals, such as shrimp, crabs, and starfish, can suffocate and die immediately when covered with oil. MAMMALS and BIRDS are affected when oil gets into their fur and feathers, robbing them of valuable insulation. Animals higher up the FOOD CHAIN are also poisoned by eating oil-soaked prey. Some animals even eat the oil itself.

There are now laws that help protect against oil pollution in the oceans. MARPOL, the International Convention for the Prevention of Pollution from Ships, prohibits the discharge of oil in oceans and coastal waters. The United States has strengthened its laws against oil pollution by enacting the MARINE PROTECTION, RESEARCH, AND SANCTUARIES ACT. Conservationists are pressuring Congress to pass a law that would require all oil tankers arriving in U.S. waters to have double hulls as added protection against accidental OIL SPILLS. [See also BIOREMEDIATION; EXXON VALDEZ; MARINE POLLUTION; OCEAN DUMPING; and OIL DRILLING.]

involve injecting fluids, such as water or steam or a combination of water and other chemicals into the oil-bearing rocks, forcing oil into the recovery pipe. Tertiary recovery methods are expensive, however, and are not often used. [See also OIL POLLUTION; OIL SHALE; OIL SPILLS; and PETROLEUM.]

Oil Pollution

▎The release of oil into the ENVIRONMENT. Oil pollution is a particular problem in the world's OCEANS. It is estimated that more than 6.5 billion pounds (3 billion kilograms) of oil contaminate Earth's seas every year.

◆ Simply shoveling oil up from the sand is one of the physical processes used to clean up spilled oil.

Oil Shale

▶ **F**ine-grained rock that contains *kerogen*—a solid, waxy, substance that can be used as a FUEL. Oil shale is removed from the ground by surface or subsurface MINING. Once removed, the shale rock is crushed and heated, changing the kerogen contained in the rock into a gas. In time, a thick, brown liquid known as shale oil is produced.

Like other FOSSIL FUELS, shale oil is burned to produce energy. Colorado, Utah, and Wyoming contain the world's largest known deposits of oil shale. Significant shale deposits are also located in Canada, China, and the former Soviet Union.

Scientists estimate that the heavy oil produced from shale deposits could supply the United States with energy for many years to come. However, manufacturing oil from shale rock is expensive. It takes the energy equivalent of one-third of a barrel of crude oil to mine and purify one barrel of shale oil. Thus, oil shale is not a practical fuel source in terms of economics.

Oil shale use and production can also damage the ENVIRONMENT. Like other fossil fuels, when shale oil is burned for energy, CARBON DIOXIDE, NITROGEN OXIDES, and PARTICULATES are released into the ATMOSPHERE. These primary pollutants are known to contribute to several environmental problems, including ACID RAIN and the GREENHOUSE EFFECT. In addition, huge amounts of water are needed to process shale oil rock. This can become a problem in the dry areas of the country, where the largest shale deposits exist. Scientists experimented with various techniques for processing oil shale without damaging the environment during the early 1980s. These projects were abandoned in 1986 because of excessive costs. [*See also* AIR POLLUTION; PETROCHEMICAL; PETROLEUM; and SURFACE MINING CONTROL AND RECOVERY ACT (SMCRA).]

◆ Oil shale has the potential of being an important source of energy. Today, however, producing fuel from oil shale is much too expensive.

Oil Spill

▶ **A**ny spillage, on land or at sea, of PETROLEUM or its refined products. Oil spills are among the worst environmental disasters because oil is poisonous to almost all living organisms.

Large spills on land usually are the results of well blowouts or pipeline ruptures caused by faulty pipe seam welds or plumbing equipment. More unusual causes include earthquakes and acts of war. There is usually less spillage from pipelines than occurs in accidents involving ships or offshore platforms because spill sensors and mechanisms for shutting down sections of pipeline are widely used. Unless a spill reaches a watercourse, it is relatively contained by the land itself, which restricts the spreading of oil.

During the Gulf War in 1991, enormous spillage occurred on land and sea. The Iraqi military forces sabotaged and ignited more than 700 Kuwaiti production wells. The Iraqis also caused the world's largest **maritime** spill. Iraqi forces deliberately released more than half a million tons (450,000 metric tons) of petroleum into the Persian Gulf from tankers and an offshore facility at Kuwait's Sea Island Terminal.

Oil spills occur frequently in wartime. During the Iran-Iraq war of 1981–89, more than 300 attacks were made on tankers. Off the Atlantic coast of the United States during World War II, German submarines sank tankers carrying a total of more than 400,000 tons (360,000 metric tons) of oil and refined

◆ Applying chemical dispersants to oil spills will break up oil and allow it to sink, but dispersants do not reduce the toxicity of oil.

products. During the same war, American submarines in the Pacific attacked Japan's merchant fleet with such success that petroleum loaded aboard Japanese ships in the Dutch East Indies (now Indonesia) was more likely to be lost at sea than to reach Japan.

Maritime accidents rather than acts of war have been chiefly responsible for raising public concern over oil spills. In 1967, the tanker *Torrey Canyon* foundered and spilled more than 100,000 tons of petroleum into the English Channel. Similar disasters occurred when the *Amoco Cadiz* ran aground off the French coast in 1978, and when the EXXON VALDEZ ran aground in Prince William Sound in Alaska in 1989.

Offshore blowouts are also a concern. Notable blowouts include the 1969 Santa Barbara spill that fouled the California coastline; the 1977 North Sea spill; and the 1979

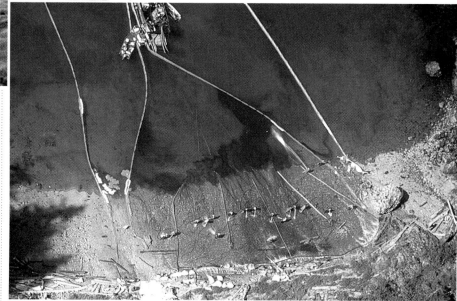

◆ Oil spills in water may be partially contained by floating barriers.

Oil Spill 95

◆ Oil from the *Exxon Valdez* is being skimmed off the surface of Prince William Sound, Alaska, by an Egmopol belt skimmer.

Gulf of Mexico spill, the largest accidental spill to date—over half a million tons of oil. Large-scale spills are responsible for only about 5% of the oil polluting the OCEANS. Most come from small accidents, leaks, and even standard operating practices, such as discharging **oil bilge washings** from tankers.

CLEANUPS

Methods exist for cleaning up maritime spills, but they are not always effective. Containment booms can keep oil from spreading on water only if deployed soon after a spill occurs. Chemical dispersants alter water's surface tension so that oil sinks, but the oil remains toxic, and some of the dispersants themselves may be harmful to ECOSYSTEMS. Researchers are investigating the

potential of nonpolluting, "oil-eating" BACTERIA capable of breaking down petroleum HYDROCARBONS.

COSTS

The cost of maritime spills is high in both ecological and economic terms. The *Amoco Cadiz* spill contaminated more than 130 beaches, killed 30,000 seabirds and 230,000 tons of FISH, crabs, and lobsters, destroyed valuable oyster and seaweed beds, and generated lawsuits totaling $750 million, filed on behalf of 400,000 local inhabitants whose livelihoods had been affected. The havoc wreaked by a spill may be very long-lived. In 1989, a ship loaded with diesel fuel sank off the Western ANTARCTICA peninsula, and the resulting oil slick threatened nearby Litchfield Island's unique

WILDLIFE with EXTINCTION. The spill's effects may still be evident in the year 2100, because diesel fuel takes vastly longer to break down in sub-zero Antarctic conditions than it does in warmer regions.

SPILLS ON THE DECLINE

In heavily industrialized areas, spills can occur almost daily—the number of incidents in the New York-New Jersey harbor, for example, peaked at 275 in 1989. It has been estimated that from three million to more than six million tons of petroleum hydrocarbons were spilled into the world's oceans every year during the 1970s and early 1980s. The amount of spillage decreased sharply during the 1980s, dropping from about one and a half million tons in 1981 to a little more than half a million tons in 1989. This is due partly to the decline of certain wasteful practices, such as discharging oil bilge washings at sea, and partly to a growing body of laws regulating or prohibiting POLLUTION of the seas.

International pressure is mounting to require supertankers to have double hulls as an added protection against oil spills. This measure was first proposed during the 1960s but successfully opposed by the oil companies. [*See also* CHEMICAL SPILLS; MARINE POLLUTION; and OIL DRILLING.]

Old-Growth Forest

▐ Any mature or ancient FOREST, between 100 and 500 years old, depending on type, that has never been harvested. In the United States, most old-growth forests are found in the Cascade Range of northern California, western Oregon and Washington, southeast Alaska, and worldwide.

In the United States, about 2.3 million acres (930,810 hectares) of old-growth forest remain, but less than one million acres (400,000 hectares) are designated as WILDERNESS areas and thus are protected from logging. In the past few years, the U.S. FOREST SERVICE has allowed 60,000 acres (24,000 hectares) of 200-year-old trees to be cut down annually. Douglas fir, western hemlock, giant sequoia, redwoods, loblolly pine, cedar, and Sitka spruce can be found in old-growth forests.

Heavy cutting of old-growth forests threatens the NORTHERN SPOTTED OWL, tree voles, marbled murrelets, and dozens of other SPECIES dependent on this HABITAT. Excessive cutting of old-growth forests in the past century has also devastated valuable SALMON runs in Washington, Oregon, and California. Many environmentalists feel that all remaining old-growth forests should be protected and preserved as natural monuments.

◆ Fewer than half of the remaining old-growth forests in the United States are in protected wilderness areas.

◆ The omnivorous raccoon eats a wide variety of foods, which helps it thrive not only in the wild, but also in towns and cities.

◆ Because bears are omnivores, they will eat fish, acorns, berries, fruit, nuts, honey, and plant leaves and roots.

Omnivore

An animal that eats both PLANTS and other animals. Whereas some animals eat only plants or only other animals, omnivores eat some of both.

Human beings are mostly omnivores, eating a variety of grains; fruits; vegetables; and animals, such as cows, chickens, pigs, and FISH. Some animals that many people think are CARNIVORES are actually omnivores, such as black bears and GRIZZLY BEARS.

Being an omnivore allows an animal to adapt to a wide range of food sources. Many of the animals that have been most successful at surviving in close proximity to dense human populations are omnivores. In the United States, raccoons, skunks, coyotes, opossums, and crows are examples of omnivores. Because they can eat a wide variety of foods, including GARBAGE thrown out by people, these omnivores can adapt to living in towns and cities. [*See also* CONSUMER; FOOD CHAIN; HERBIVORE; PREDATOR; and TROPHIC LEVEL.]

Open-Pit Mining

A MINING method in which heavy equipment is used to dig large holes in the ground and expose ore. This method of mining is used when the ore deposit lies near or on

◆ The Bingham Canyon copper mine in Utah was used to remove large deposits of the ore lying near the surface. The copper was removed from horizontal layers, which then were cut away, leaving the periphery to serve as roads.

the surface and extends only a few hundred feet (less than 100 meters) below the surface. The world's largest mines are open-pit mines. In general, open-pit mines are less expensive and less dangerous to miners than underground mining operations; however, open-pit mines result in HABITAT destruction.

The Bingham Canyon COPPER mine in Utah is the world's largest open-pit mine—the largest artificially made hole in history, in fact, and likely to grow ever larger as miners dig ever deeper for copper ore. The mine produces about 250,000 tons (about 227,000 metric tons) of copper each year. For every metric ton of copper produced, about 1,320 tons (about 1,200 metric tons) of waste material, much of it dangerously toxic, are

also produced. Laws have been passed to protect the ENVIRONMENT in areas where open-pit mining is taking place. [*See also* MINERALS; MINING; and TAILINGS.]

Organic Farming

❱A n agricultural technique that uses our understanding of nature as a guide for gardening. In addition to protecting the ENVIRONMENT, organic farming saves money by reducing the need for IRRIGATION, PESTICIDES, and synthetic fertilizers.

In natural ECOSYSTEMS, when PLANTS and animals die, BACTERIA,

FUNGI, and other DECOMPOSERS consume their bodies, releasing locked-up nutrients in the process. Organic farmers try to replicate this process by adding manure, compost, and other organic matter to the soil, rather than adding synthetic fertilizers. Organic farmers also encourage soil fertility by rotating crops. They try to avoid soil EROSION by keeping the ground planted at all times. Rather than synthetic pesticides, organic farmers use a system called INTEGRATED PEST MANAGEMENT (IPM) to control INSECTS, FUNGI, and other pests that eat plants.

HISTORICAL DEVELOPMENT OF ORGANIC FARMING

The historical roots of organic farming are difficult to trace. Before synthetic pesticides and fertilizers were developed in the 1800s, all farming was organic. However, large-scale organic farming can be traced to at least the 1940s, when farmers began to investigate the importance of HUMUS in soil. There was an explosion of scientific publications during that decade on the topic of organic farming and gardening, including J. I. Rodale's *Organic Farming and Gardening* in 1942, and Sir Albert Howard's *An Agricultural Testament* in 1940. Most authors concluded that farms and gardens should be treated like living natural ecosystems in which nutrients are recycled naturally.

Soon after, with the publication of Rachel CARSON's *Silent Spring* in 1954, farmers began to worry about the effects of synthetic fertilizers and pesticides on the environment. A small segment of farmers began to adopt organic methods. How-

◆ Organic farming uses a variety of techniques to fertilize soil and control pests naturally. A farmer spreads animal manure to maintain soil fertility.

ever, for many years, the organic method was considered unusual and generally ignored by commercial farmers.

By the 1980s, more farmers and scientists began asking questions about organic farming. In 1988, organic farming received an important boost when the U.S. DEPARTMENT OF AGRICULTURE announced a program to fund research on SUSTAINABLE AGRICULTURE. The goal of this program is to lower the costs of crop production and increase profits while protecting the health of the environment.

In 1990, the federal govern-ment became more involved with organic farming when Congress passed the Organic Foods Pro-duction Act as part of the 1990 Farm Bill. According to this act, as of October 1993, food sold as "organic" has to meet requirements set by the U.S. Department of Agriculture.

IMPORTANT ORGANIC FARMING TECHNIQUES

Organic farmers use different tech-niques to maintain soil fertility and control pests without using poten-tially harmful chemicals.

Crop Rotation

CROP ROTATION is a farming practice that helps keep soil fertile through the alternating, or rotating, of differ-ent crops. It is well known that some plants can rapidly deplete nutrients from the soil, whereas others add nutrients. For instance, corn, tobacco, and cotton are crops that can quickly rob nutrients from the soil, particularly nitrogen. On the other hand, LEGUMES, such as oats, rye, and barley, add nitrogen to the soil. In crop rotation, these different types of crops are alter-nated each season to keep soils fertile. Crop rotation also helps reduce infestations by insects because populations are not able to grow large over the years.

Organic Fertilizers

Adding organic matter—various forms of living or dead plant and animal material—to soil is another important technique of organic farming. Adding animal manure is probably the most widely used method for maintaining soil fertility. Manure is an important source of nitrogen, potassium, and phosphorus and is also highly rich in bacteria.

Integrated Pest Management

Integrated pest management is a system of pest control that combines a variety of biological, cultural, and physical control techniques. Biological pest controls are methods that use living organisms to fight pests. Among the more popular techniques is the use of insecticide sprays containing bacteria. When sprayed on infested plants, these INSECTICIDES cause insects to get sick and die. Other methods involve releasing, attracting, and protecting natural predators of insect and rodent pests to control pest populations.

Cultural control has to do with the way crops are planted. For instance, properly spacing plants helps to reduce pest problems because crowded plants are usually weak and more prone to disease and insect damage. Other cultural control methods include planting in fertile and moist soils to keep plants healthy and using insect-resistant strains of plants.

Physical controls are techniques that try to keep pests from reaching plants. The use of barriers is the most popular physical control technique. Barriers can be extremely effective because they stop pests from reaching plants in the first place. For instance, tree bands, which are strips of cloth or cardboard wrapped around the trunk of a tree, prevent nonflying insect pests, such as snails, slugs, ants, and GYPSY MOTH caterpillars, from devouring the leaves of the tree. Other common methods include the use of sticky traps and layers of fine dust to catch pests. [*See also* AGROECOLOGY; BIOACCUMULATION; COMPOSTING; DDT; FUNGICIDE; NITROGEN FIXING; NO-TILL AGRICULTURE; PEST CONTROL; PCBS; RODENTICIDE; and SOIL CONSERVATION.]

Organization of Petroleum Exporting Countries (OPEC)

Association of 13 nations that rely mainly on exporting oil for income and whose members unite in an effort to increase their profits from oil sold in the international marketplace. The nations of OPEC—Algeria, Ecuador, Gabon, Indonesia, Iran, Iraq, Kuwait, Libya, Nigeria, Qatar, Saudi Arabia, the United Arab Emirates, and Venezuela—control between two-thirds and three-fourths of the world's recoverable oil, and produce about one-third of the oil

◆ Supertankers that transport oil from OPEC nations run the risk of accidents, such as massive oil spills that pollute the environment and kill or endanger water animals and land animals.

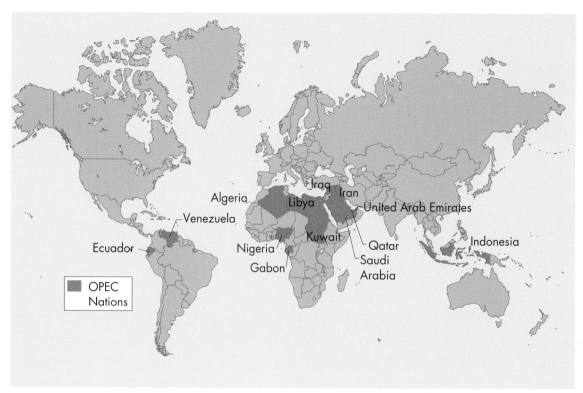

◆ Huge oil reserves are controlled by the 13 members of OPEC.

Map labels: Algeria, Iraq, Iran, Libya, United Arab Emirates, Venezuela, Nigeria, Kuwait, Qatar, Indonesia, Ecuador, Gabon, Saudi Arabia

OPEC Nations

traded on the open market. OPEC is made up of four official sections: the OPEC Conference sets prevailing policies; the Board of Governors puts policies into effect and controls the Secretariat, which deals with managerial and diplomatic affairs; and the Economic Commission recommends oil prices.

OPEC was founded by Iran, Iraq, Kuwait, Saudi Arabia, and Venezuela in 1960, when the U.S. and European oil companies controlled the PETROLEUM industry in these oil-producing countries. Each oil company paid the local government taxes and a royalty—share of the profit—based on the going price of crude oil on the world markets. When world oil production far surpassed demand in 1959 and 1960, some oil companies lowered

the going rate and consequently the amount they paid the local governments. In response to the price cut, OPEC was set up so that the oil-producing nations could gain more control over oil production and pricing.

During the 1960s, oil supplies kept up with the demand, so OPEC had very little influence on world oil prices. But in the 1970s, oil demand exceeded supplies available from non-OPEC sources, so OPEC raised its prices sharply. Once again in the 1980s, oil supplies surpassed demand and OPEC had to lower the price of oil. It also set output limits for all OPEC members, except Saudi Arabia. Again in 1986, OPEC cut oil prices.

Oil, a FOSSIL FUEL, is a NON-RENEWABLE resource so it is impor-

tant to conserve our oil reserves. But during recovery from underground wells or transport by pipeline or supertanker, oil can burn or spill, creating environmental hazards, as evidenced by the EXXON VALDEZ oil spill disaster in Prince William Sound and the smoke from burning oil wells on the Persian Gulf. Engineers are searching for more efficient ways to tap into underground oil sources and better ways to transport oil so that the ENVIRONMENT is protected. Many environmental experts hope more people will opt for ALTERNATIVE ENERGY SOURCES, such as solar power, thermal power, and WIND POWER, that can provide the same amount of ENERGY as oil without threats to the environment. [*See also* FUELS; OIL DRILLING; and OIL SPILL.]

◆ Overgrazing strips the land of vegetation. As a result, the soil is no longer protected from wind and water erosion.

Overgrazing

▶The practice of permitting LIVESTOCK to graze in one area until all vegetation covering the ground is destroyed. Overgrazing leaves the SOIL unprotected against water and wind EROSION. It also exposes the bare soil to excess sunlight, which can chemically alter its composition.

In the past, GRAZING animals roamed freely on vast plains, eating grass in one area and then moving on to another. The grass had time to grow back. Then people began to confine their livestock behind fences. Often, too many animals grazed in one area. Overgrazing usually occurs in overpopulated regions, where it is necessary to raise as many livestock as possible to feed the people.

Overgrazing can destroy not only GRASSLANDS but neighboring FORESTS, as well. In parts of India, for example, goat herders hack off tree limbs to feed their goats because the animals have stripped the soil of its vegetation. The barren trees die.

Overgrazing can also cause rivers and streams to overflow. With no vegetation to hold water in the soil, rainwater flows down into waterways instead of soaking into the ground. The overflowing rivers and streams cause more damage to the land.

A fertile grassland can become DESERT if it is overgrazed. Grain once grew abundantly in North Africa's ancient Carthage, but now the land can no longer support farming or the raising of livestock. Because of overgrazing and overfarming, most of the region is now part of the Sahara Desert.

What can be done to prevent overgrazing? Livestock owners can begin to control the number of animals they graze in one area. They can rotate their herds among grazing areas to allow time for new plant growth. Another possible solution is to provide supplementary food to livestock so that they need less grazing land. [*See also* DESERTIFICATION.]

Overpopulation

▌▶General term that refers to high abundance, or excess, of a single SPECIES that may exceed the environment's CARRYING CAPACITY, or ability to support it. Rapid POPULATION GROWTH (a so-called population explosion) can lead to large populations of PLANTS, animals, or microbes that may not be sustained by the local ENVIRONMENT. Populations that cannot be supported by the available food, water, and space in the environment are subject to high death rates from disease and starvation.

Sometimes EXOTIC SPECIES can overpopulate an area because they outcompete an existing species, or the area may have no PREDATORS or diseases capable of controlling the species' population. Overpopulation can also refer to human settlements with high population densities. Large cities can be seen as overpopulated because they can only be supported by transferring large amounts of food and energy from other, less populated areas. Some countries might be seen as overpopulated for similar reasons; that is, the country cannot produce enough food energy to sustain its population.

◆ In India, many people suffer from malnutrition as a result of overpopulation.

◆ Some environmentalists consider developed countries to be overpopulated because these people use more of the world's resources than do people of developing countries.

Oxygen

▶ Nonmetallic chemical element that exists as a gas at room temperature. Oxygen, an odorless, tasteless, and colorless gas, is a common element in Earth's crust. In combination with other chemicals, it makes up approximately 25% of every living thing. It also makes up about 21% of the volume of Earth's ATMOSPHERE.

Most oxygen in the air is released as a waste product of PHOTOSYNTHESIS. Photosynthesis is the chemical process by which PLANTS and most other PRODUCERS use sunlight, CARBON DIOXIDE, and MINERALS from the SOIL to make food in the form of sugar. The oxygen in the air is used by most organisms in RESPIRATION. Respiration is a chemical process that allows organisms to obtain energy from food. In terms of the raw materials involved, respiration is the opposite process of photosynthesis.

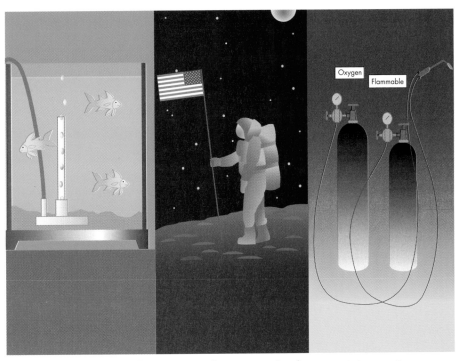

◆ In environments where there is an inadequate supply of oxygen, it is sometimes provided with oxygen pumps or tanks.

Many animals take in oxygen through specialized organs, such as lungs and gills. INVERTEBRATES such as sponges, flatworms, and some INSECTS take in oxygen directly through their cells. Plants take in oxygen through structures called *stomata* located in their leaves. Once the oxygen circulates through its cells, organisms expel carbon dioxide. The carbon dioxide is used by producers in a never-ending gas-exchange cycle.

OTHER USES OF OXYGEN

Besides being used by organisms, oxygen is essential for the burning of wood, COAL, and other FUELS. French scientist Antoine Laurent Lavoisier noted that when a material burned, it combined with an element in air. In 1779, he named that element *oxygen.*

Oxygen is used in large quantities for industrial processes such as the smelting and refining of iron and steel. Combined with acetylene, oxygen is used in torches

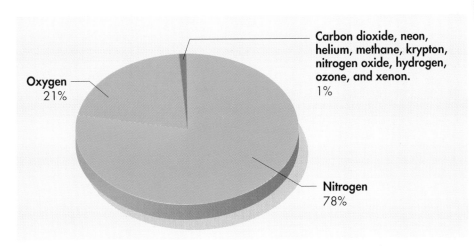

◆ Approximately 21% of the air we breathe is oxygen gas. Air is mostly nitrogen gas, which animals breathe in but do not use.

for cutting and welding metals. Oxygen mixtures are breathed by some hospital patients, deep-sea divers, and high-altitude pilots. Liquid oxygen, which is obtained by cooling oxygen gas, is used in rocket fuels.

OXYGEN AND THE ENVIRONMENT

Oxygen has two allotropes—distinct forms that differ in structure. The oxygen used by organisms has the formula O_2 and is called oxygen. OZONE (O_3) is a form of oxygen made up of three atoms. Both oxygen and ozone are needed in our ENVIRONMENT. The ozone layer, for example, protects us from ULTRAVIOLET RADIATION from the sun.

The amount of oxygen in our atmosphere remains constant because of photosynthesis. However, human activities can alter the levels of both forms of oxygen in the atmosphere. When FORESTS are cut down or burned to make room for roads, farms, or factories, the atmosphere is deprived of the oxygen it normally receives through photosynthesis. If the trees are burned, the damage is worsened because burning wood emits a great deal of carbon dioxide. Scientists believe the buildup of carbon dioxide in the atmosphere may lead to GLOBAL WARMING.

Chlorofluorocarbons (CFCs), used in AEROSOL cans and as air conditioner and refrigerator coolants, drift up into the atmosphere. When they reach the ozone layer, the chlorine in these compounds is released as the CFCs react to the sun's ultraviolet rays. The chlorine

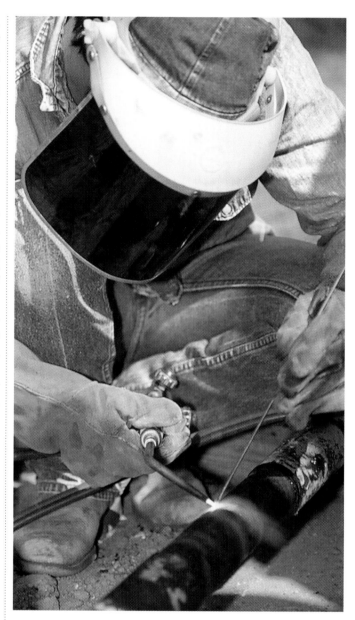

◆ Oxygen, which combines with nearly all other elements, is used for different purposes, such as to cut and weld metal.

combines with oxygen in the ozone layer, breaking apart the ozone into oxygen atoms. This destroys the ozone layer. Scientists have discovered a hole in the ozone layer that is allowing more ultraviolet radiation to reach Earth. In 1990, the United States and 92 other nations agreed to stop production of CFCs and other ozone-destroying chemicals by the twenty-first century. However, some scientists believe the effects of CFCs already in the atmosphere could last for hundreds of years. [*See also* CLEAR-CUTTING; CLIMATE CHANGE; DISSOLVED OXYGEN; GREENHOUSE EFFECT; and OXYGEN CYCLE.]

Oxygen Cycle

▌The cycling of the chemical OXYGEN among organisms, land, water, and air. Oxygen, like all other elements, moves from one place to another in the land, or LITHOSPHERE; the water, or HYDROSPHERE; and the air, or ATMOSPHERE, in a BIOGEOCHEMICAL CYCLE. Oxygen is chemically very reactive and is found in many different compounds. It is found in such compounds as water (H_2O), CARBON MONOXIDE (CO), CARBON DIOXIDE (CO_2), SULFUR DIOXIDE (SO_2), and nitrate (NO_3). It is also found in MINERALS like limestone ($CaCO_3$) and iron ore (Fe_2O_3). These compounds link the oxygen cycle with

◆ Whether on land or in bodies of water, plants produce oxygen in photosynthesis. The oxygen comes from breaking apart water into its two elements, hydrogen and oxygen.

◆ Oxygen in various forms cycles through organisms, air, water, and land. Activities of humans are part of this cycle.

cycles of hydrogen, CARBON, sulfur, calcium, and iron.

OXYGEN IN THE ATMOSPHERE AND IN WATER

Water vapor in the ATMOSPHERE is broken apart by ULTRAVIOLET RADIATION, releasing oxygen. Reactions occur in which oxygen produces OZONE, which in turn breaks down into oxygen gas (O_2) and atomic oxygen (O). Oxygen gas in the atmosphere enters and leaves bodies of water and soil pores. The colder the water, the more oxygen it will hold.

BACTERIA AFFECT OXYGEN COMPOUNDS

Oxygen gas combines with nitrogen gas N_2 to form nitrate. This reaction is part of the nitrogen cycle and is called *nitrogen fixation*. It occurs in the soil through the action of bacteria in the process called NITROGEN FIXING. The nitrogen in nitrates then is turned into nitrogen gas with the help of denitrifying bacteria in the soil.

PHOTOSYNTHESIS AND RESPIRATION

Plants carry out two kinds of reactions—photosynthesis and respiration—which cycle oxygen. In PHOTOSYNTHESIS, carbon dioxide, sunlight, and water undergo a process in which solar energy is turned into chemical energy and stored in sugar ($C_6H_{12}O_6$). The carbon and oxygen in sugar come from carbon dioxide. The water is split into hydrogen and oxygen and

the oxygen is released into the air. In respiration, oxygen from the air combines with hydrogen and is turned back into water. Thus, the oxygen cycle is completed. Potosynthesis and respiration are important parts of the oxygen cycle.

Energy stored in plants in the form of sugar is passed along to animals that eat the plants. Animals use plant sugar as a source of energy for their own energy needs. Animals that eat other animals also use sugar for energy. Thus, sugar is transferred to CONSUMERS in the FOOD CHAIN.

When organisms use sugar they break it down by respiration into released energy, carbon dioxide, and water. Plants use carbon dioxide and water to make sugar. This is how oxygen in carbon dioxide is cycled from organisms to plants. Organisms use oxygen gas to bond with hydrogen to form water. In this way, oxygen cycles from plants to all organisms.

Ozone

▶A form of OXYGEN gas that contains three oxygen atoms (O_3) rather than two (O_2). Small amounts of ozone occur naturally about 52,000 feet (about 15,000 meters) above Earth's surface in the region of the ATMOSPHERE called the STRATOSPHERE. In the stratosphere, ozone forms a layer about 18 miles (about 29 kilometers) thick. This layer, called the OZONE LAYER,

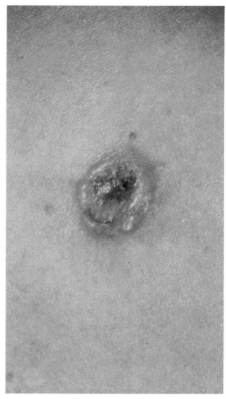
◆ Overexposure to ultraviolet radiation can cause skin cancer.

absorbs almost all of the ULTRAVIOLET RADIATION generated by the sun.

The increased amounts of ultraviolet radiation that reach Earth can be extremely harmful to living things. Overexposure to ultraviolet radiation can cause skin CANCER and mutations in a cell's deoxyribonucleic acid ·(DNA). It is also known to reduce crop yields and kill the tiny organisms called PLANKTON that populate Earth's OCEANS. Many living species could not exist without the protection from ultraviolet radiation that is provided by the ozone layer.

Ozone in the atmosphere can be harmful. Near Earth's surface, ozone is a secondary pollutant, that is, a major component of SMOG.

Ozone is created when AUTOMOBILE and factory emissions react with solar RADIATION and other air pollutants. In small amounts, ozone is an irritant to the eyes and lungs. In large amounts, ozone can be highly toxic to humans. Because ozone can destroy cells, it is used in some SEWAGE TREATMENT PLANTS to kill BACTERIA.

Many environmentalists are very concerned about the high levels of ozone that occur in many large cities. Cities often issue "ozone alerts" to warn people when ozone levels are high. During such alerts, citizens are usually asked to carpool and to reduce highway speeds. Both of these activities help reduce the emissions that raise ozone levels.

Many scientists are also con-cerned about the use of some human-made chemicals, particularly chlorofluorocarbons (CFCS), that are contributing to the destruction of the protective ozone layer. The United States has already banned the use of AEROSOLS in spray cans, one major source of CFCs. By the year 2000, use of other substances that pose a significant threat to the ozone layer will also be banned. One of these substances is the FREON used in refrigeration and air conditioning systems in both homes and automobiles. In addition to bans on CFCs issued by the United States, many other industrialized nations have also agreed to reduce the manufacture and eliminate CFCs by the year 2000. [*See also* AIR POLLUTION; CARCINOGEN; and OZONE HOLE.]

Some Effects of an Increase in UV Radiation

1. Increase in nonmelanoma—disfiguring but not usually fatal

2. Increase in melanoma skin cancer—often fatal

3. Increase in eye cataracts

4. Severe sunburn

5. Suppression of immune system—weakened defenses against diseases

6. Increase in eye-burning photochemical smog

7. Increase in the deposit of acid near areas where fossil fuels are burned

8. Increase in eye cancer in cattle

9. Damage to many species of land plants, including food plants for humans

10. Damage to aquatic plant species essential to food chains

11. Degradation of plastics

Ozone Hole

❚❚❚The area of thinning in the OZONE LAYER above ANTARCTICA. This region of thinning, or "hole," discovered in the 1980s, focused worldwide attention on the relationship between chlorofluorocarbons (CFCS) and ozone depletion.

Public awareness of OZONE began in the 1970s when scientists revealed that human-made chemicals known as CFCs could contribute to thinning of the OZONE LAYER. This layer, found about 50,000 feet (about 15,000 meters) up in Earth's ATMOSPHERE, is critical to life on Earth. By absorbing the harmful ULTRAVIOLET RADIATION given off by the sun, the ozone layer acts as a sunscreen for the Earth.

The public was justifiably concerned. Ultraviolet light is very dangerous to humans, causing skin CANCER, eye damage, and other medical problems. If the amount of ozone in the STRATOSPHERE, the upper portion of the atmosphere, were to decrease, more ultraviolet light would be able to pass through the atmosphere and reach Earth's surface. In 1978, the U.S. government responded to public pressure and banned the use of CFCs in one of its major sources—AEROSOL spray cans.

THE DISCOVERY OF THE OZONE HOLE

In 1984, a group of scientists working at Halley Bay, Antarctica, made a startling discovery. The scientists found that between the months of September and October, ozone levels in the atmosphere above the Antarctic had dropped by about 50%. The next year, the results of this research were published in the British scientific journal *Nature*. This was the first time the public had heard of the now famous ozone hole.

After news of the ozone hole had reached the scientific community, many scientists decided to look into this problem. NASA scientists, for instance, reviewed information that had been collected by the weather satellite *Nimbus 7*. The satellite data confirmed that an ozone hole

◆ Evidence of a hole in the ozone layer over the North Pole was gathered by equipment in a DC-8 flown by NASA. A sample flight path is shown, with the darkest pink area indicating the likely position of the hole.

was present over the Antarctic as early as 1979.

More recent studies have shown that the Antarctic ozone hole forms during the winter months and closes during the summer. Scientists suggest that this may be due to the presence of ice CLOUDS and other special conditions that exist only in the Antarctic. New evidence has shown

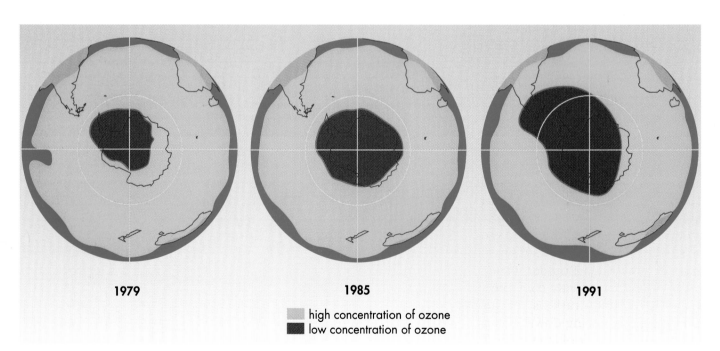

| 1979 | 1985 | 1991 |

◻ high concentration of ozone
◼ low concentration of ozone

◆ This series of images gathered by NASA's *Nimbus 7* weather satellite shows the ozone hole, at the South Pole, identified by the purple color.

that ozone thinning is now taking place over the Arctic as well. The discovery of the ozone hole over Antarctica raised public awareness and focused attention on the problem of ozone depletion.

PROTECTING THE OZONE LAYER

CFCs are commonly used as propellants in aerosol spray cans. But they are also extensively used in air conditioners and refrigerators, and in the manufacture of plastic foam products. Ever since the U.S. ban on spray cans containing CFCs in 1978, environmentalists have been searching for ways to reduce and eliminate the manufacture of these harmful chemicals. The MONTREAL PROTOCOL, a 1987 international conference in Montreal, was one of the first meetings held to discuss the problem of ozone depletion. At this conference, industrialized nations agreed to cut the manufacture of CFCs in half by the year 2000. This agreement was later amended to state that CFCs and other harmful substances must be totally eliminated by the year 2000.

Recent studies have shown that agreements to protect the ozone layer are having positive effects. The worldwide production of CFCs has been greatly reduced, and the search is on for safer chemical substitutes. Scientists at the NATIONAL OCEANIC AND ATMOSPHERIC ADMINISTRATION (NOAA) in Boulder, Colorado, have also recently confirmed that CFC concentrations in the atmosphere are going down. Some scientists are even looking for ways to replenish ozone levels in the atmosphere. [*See also* OZONE.]

Ozone Layer

Layer of OZONE gas (O$_3$) consisting of three oxygen atoms located in the upper ATMOSPHERE that shields Earth from dangerous ULTRAVIOLET RADIATION emitted by the sun. The ozone layer is located within the part of Earth's atmosphere called the STRATOSPHERE. The stratosphere is one of the four layers that make up Earth's atmosphere. The OZONE lies about 50,000 feet (15,000 meters) above Earth's surface. This protective layer acts as a sunscreen for the many forms of life on Earth by absorbing the damaging ultraviolet radiation given off by the sun. Ultraviolet radiation can be extremely dangerous to life. With-

out the ozone layer, humans and other organisms would be constantly bombarded with ultraviolet radiation and risk the development of many diseases. In humans, these diseases include skin CANCER, eye damage, and other medical problems. For example, ultraviolet radiation is known to interfere with PHOTOSYNTHESIS and cause damage to crops and the tiny marine organisms known as PHYTOPLANKTON.

Today, scientists are concerned that the use of chemicals called chlorofluorocarbons (CFCs) may be severely damaging the ozone layer. CFCs are used extensively in refrigerators, air conditioners, and in the manufacture of foam products such as styrofoam. CFCs are also used as **propellants** in AEROSOL spray cans. CFCs contribute to ozone depletion

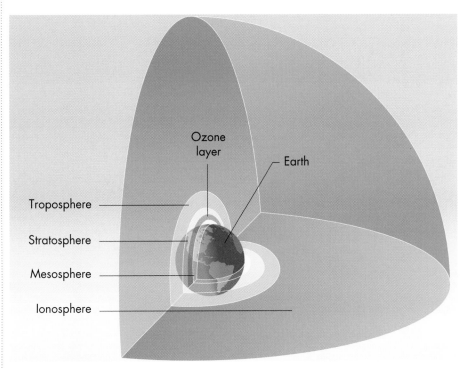

◆ The ozone layer is located in the upper atmosphere, known as the stratosphere, about 50,000 feet (15,000 meters) above Earth's surface.

after they drift up into the ozone layer in the upper atmosphere. When CFCs are released into the air, they gradually rise into the atmosphere. In time, CFCs are broken apart by the energy contained in sunlight. As ozone molecules are broken apart, many of the oxygen atoms recombine to form **diatomic** oxygen gas. As the OZONE breaks apart, a decrease in the total amount of ozone takes place.

Scientists are very concerned about ozone depletion because the ozone layer is thinning faster than it forms by natural processes. In 1984, researchers working at Halley Bay on the coast of Antarctica made a startling discovery. They observed a hole in the ozone layer. In 1985, scientists reported in the British scientific journal *Nature* that ozone levels in the atmosphere above Antarctica had decreased by more than 50%. NASA scientists later reviewed data from their *Nimbus 7* weather satellite that confirmed that this OZONE HOLE was present as early as 1979.

As scientists became more concerned about these data, public awareness about the thinning ozone layer grew. In 1978, the U.S. government banned the use of CFCs in aerosol products. In a conference held in 1992 about the problem of ozone depletion, 93 countries agreed to significantly reduce the manufacture and use of CFCs and other hazardous chemicals by the year 2000. One form of this reduction has already begun taking place in air conditioning and refrigeration systems used in homes, AUTOMOBILES, and industry. These products are replacing FREON, a CFC once used as a coolant, with non-CFC chemicals. Although these measures are expected to help the ozone layer recover, the process will be slow because CFCs may remain in the upper atmosphere for up to 100 years. [*See also* CARCINOGEN.]

Ozone Pollution

The buildup of OZONE near the surface of the Earth. Ground-level ozone pollution is a serious AIR POLLUTION problem throughout the industrialized world. It poses threats to both human health and vegetation.

FORMATION OF GROUND-LEVEL OZONE

Ozone (O_3) is a chemical cousin of OXYGEN (O_2) composed of three atoms of oxygen rather than the usual two. In the STRATOSPHERE, a thin layer of ozone absorbs ULTRAVIOLET RADIATION coming from the sun and protects Earth's inhabitants

◆ Los Angeles has the most serious ozone pollution problem in the United States. The hot, sunny climate and the millions of motor vehicles contribute to this problem.

from its harmful effects. However, near the surface of Earth, many human activities contribute to ozone formation, where it is often referred to as ground-level ozone.

Ground-level ozone is an example of a secondary pollutant; that is, it is not released directly from human activities. Rather, ozone is formed indirectly when primary pollutants react with each other in the ENVIRONMENT.

A variety of human activities contribute to the formation of ground-level ozone. All involve the use of FOSSIL FUELS, such as COAL, oil, and NATURAL GAS. When these substances are burned for energy to run motor vehicles, power plants, and factories, two harmful substances are given off: HYDRO-CARBONS and NITROGEN OXIDES. In the presence of sunlight, these two primary pollutants react to form ground-level ozone.

ENVIRONMENTAL EFFECTS OF OZONE POLLUTION

Ozone pollution is a major air pollution problem in many large cities of the world. During the late 1970s and early 1980s, ozone pollution in Japan and Europe was 75% higher than levels considered safe by the World Health Organization. The United States and Australia exceeded these limits by 400%.

According to statistics from the COUNCIL ON ENVIRONMENTAL QUALITY and the ENVIRONMENTAL PROTECTION AGENCY (EPA), ozone pollution in the United States dropped 21% from 1980 to 1992. However, ozone pollution continues to be a serious problem in large cities like Los Angeles, New York, and Houston where it contributes to SMOG formation. Los Angeles, by far the smoggiest city in the United States, has all the conditions necessary for severe ozone pollution, including a hot, sunny CLIMATE, 13 million people, 8 million cars, and a dozen oil refineries.

The environmental and health effects of ozone pollution are many. High levels of surface-level ozone can cause eye and lung irritation. Especially vulnerable to ozone pollution are the millions of people who already suffer from **asthma**, **emphysema**, and other respiratory illnesses. Ozone can also harm crops and other vegetation by damaging plant tissues and interfering with the plant's ability to carry out PHOTOSYNTHESIS.

REDUCING OZONE POLLUTION

Recent changes to the CLEAN AIR ACT in 1990 directly addressed the problem of ozone pollution and smog. Under these laws, nearly 100 major cities in the United States are being forced to meet national smog level standards by 1999. Heavily polluted cities like New York, Los Angeles, and Baltimore must comply by 2005. Many of these cities are also being asked to reduce ozone pollution 15% by 1997, and an additional 3% each year until the law's standards are met. The recent amendments also impose new controls on pollution emitted by motor vehicles, the chief sources of hydrocarbons and nitrogen oxides. [*See also* ACID RAIN; CFCS; GREENHOUSE EFFECT; GREENHOUSE GAS; INDUSTRIAL REVOLUTION; OZONE HOLE; OZONE LAYER; and POLLUTION.]

P-Q

Chile

Paraguay

Rio de Janeiro
Sao Paulo

Brazil

Uruguay

Buenos Aires

Montevideo

Argentina

Falkland
Islands (U.K.)

◆ The pampas of South America are used largely for agricultural purposes, but they also support native species of plants and animals.

Pampas

❚Treeless plains regions located in several South American countries. Plains are geographic regions made up of large tracts of level land with few trees. In the United States, the plains regions are largely GRASSLANDS used for agricultural purposes, such as for growing crops and grazing animals.

In South America, sections of Brazil, Argentina, Peru, and Chile are made up of plains regions called *pampas*. The main PLANT in these regions is pampas grass, a

◆ The dry pampas in Argentina are used by ranchers for grazing animals.

reedlike grass that grows in clumps. Pampas grass is a tall grass that grows to heights of more than 6 feet (2 meters). Deer, known as pampas deer, are a common animal SPECIES in the pampas of Argentina and Brazil. These red-brown deer are smaller than those common to the forest regions of the northern United States.

The pampas of South America have two distinct CLIMATES: humid and dry. Each climate region has different uses. For example, the humid pampas are fertile regions used for growing crops. These regions are fairly well populated by people. The dry pampas are less populated areas used by ranchers for grazing animals, such as horses, cattle, and sheep. [*See also* BIOME; NATIVE SPECIES; and PRAIRIE.]

Parasitism

▶A feeding relationship between two species in which one benefits while the other is harmed but not usually killed. The organism that benefits in this relationship is the **parasite**. The organism that is harmed is called the **host**.

In some ways, parasitism is similar to predation, an interaction between SPECIES in which one species kills and eats another. However, the two types of relationships differ in one important way. In parasitism, the parasite slowly feeds on the cells, tissues, or fluids of its host, causing damage to the host without actually killing it.

◆ Ticks often carry disease organisms in their bodies and transfer them to the blood of their victims.

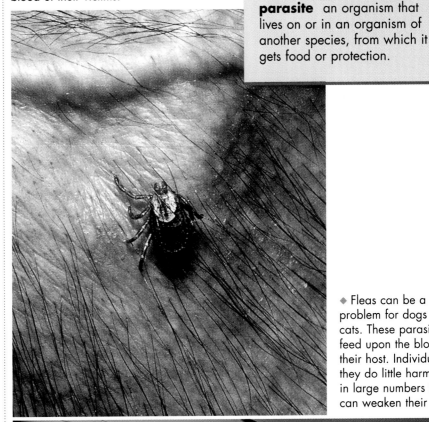

THE LANGUAGE OF THE ENVIRONMENT

host any organism that a parasitic organism lives on or in for nourishment or protection.

parasite an organism that lives on or in an organism of another species, from which it gets food or protection.

◆ Fleas can be a major problem for dogs and cats. These parasites feed upon the blood of their host. Individually, they do little harm, but in large numbers they can weaken their hosts.

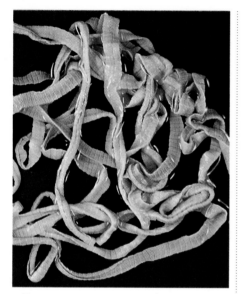

◆ Tapeworms are parasites that are sometimes found in the intestines of humans.

All types of organisms—animals, PLANTS, FUNGI, protists, and BACTERIA—have some parasitic species. The most familiar examples of parasites are the fleas, ticks, mites, and worms that live on and in dogs, cats, and other pets. Fleas and ticks, living within the fur of many MAMMALS, survive by sucking nutrient-filled blood from their hosts. Individually these types of parasites cannot kill their hosts because the damage they cause occurs slowly. However, over time and in large numbers, these parasites can weaken their hosts, making them more vulnerable to attack by other PREDATORS and disease. In addition, many parasites such as fleas and ticks can carry and spread disease. For example, the deer tick common in many FORESTS of the Untied States is a carrier of Lyme disease.

Some parasites are unique to a particular species. In such cases,

the parasite can survive only by feeding on that specific host. GORILLAS have parasites known as crab lice (*Pthirus gorillae*), that live within their fur. Although this species of lice is very similar to the crab lice that infect humans, it is specific only to gorillas. The gorilla lice feed on dead cells on the gorilla's body.

Parasitism, along with COMMENSALISM and MUTUALISM, is a type of symbiotic relationship. Symbiotic relationships are close, long-term partnerships between species. Many species of organisms, particularly large animals and plants, are hosts to parasites. Parasitism demonstrates the process of COEVOLUTION, because both the host and the parasite have evolved adaptations in response to each other. [*See also* ADAPTATION; BIOLOGICAL COMMUNITY; HEALTH AND DISEASE; and SYMBIOSIS.]

Particulates

▌Tiny, solid particles of smoke, soot, ash, dust, and other substances produced by the burning of FOSSIL FUELS. Particulates are a primary pollutant, causing nearly one-fourth of all AIR POLLUTION.

PROBLEMS FROM PARTICULATES

Particulates cause a variety of environmental and health problems. They may be the most visible form of air pollution. The emissions from AUTOMOBILES, factories, and power plants are thick and dark in appearance because of the particulates they release. These emissions often form a haze of smoke that reduces visibility. In congested cities, such

◆ The burning of diesel fuel by trucks releases clouds of particulates.

◆ When fuel is burned, the impurities are released as particulates.

as New York City and Los Angeles, long-term exposure to particulates can cause several health problems, including eye irritation and a variety of respiratory diseases.

REDUCING PARTICULATE EMISSIONS

Industry is the main source of particulates in the air. According to the U.S. CLEAN AIR ACT, factories and power plants are required to control the amounts of air pollutants they release. Many industries do this by installing devices in their smokestacks that control pollution. Known as SCRUBBERS, these devices regulate the amount of pollutants released into the air by literally "scrubbing" harmful contaminants from smoke. To do this, scrubbers use a pressurized spray of water to cleanse the incinerator smoke.

Particulates will continue to be released into the air as long as fossil fuels are used as a source of energy. However, there are several ways to reduce particulate emissions. One of the most effective ways is to reduce the use of motor vehicles, which account for nearly 20% of particulates in the air. Carpooling has also been shown to be an effective way to reduce the emission of particulates and other air pollutants. Many cities have already begun carpool programs to help reduce the amount of traffic on the roads. The use of mass transit systems instead of individual automobiles can also reduce particulate pollution. [*See also* PRIMARY POLLUTION and SMOG.]

Passenger Pigeon

▌A SPECIES of migratory BIRD in the pigeon family that was hunted to EXTINCTION by humans in the 1880s and early 1900s. Passenger pigeons were one of the most abundant types of birds in eastern North America in the 1800s. Scientists estimate that passenger pigeons once had a population size of 3 to 5 billion birds, making up 25% of all North American birds. The species was officially classified as extinct when the last known passenger pigeon, a bird named Martha, died in the Cincinnati Zoo in 1914.

Passenger pigeons were about 13 inches (33 centimeters) in length with a long pointy tail. Males were characterized by a pinkish chest, neck, and throat and a blue-gray body and head. Females were smaller and had duller colors.

Passenger pigeons inhabited mostly areas where they filled the trees with nests. The birds ate mostly acorns, seeds, and beech-nuts. Passenger pigeons migrated in extremely large flocks. Flocks

◆ The passenger pigeon (*Ectopistes migratorius*) was the most abundant type of bird in North America until it was hunted to extinction in the late 1800s.

of passenger pigeons were known to contain several million birds.

CAUSES OF EXTINCTION

The passenger pigeon population in North America declined rapidly for several reasons. As FORESTS were cut down for use as farms, timber, and FUEL, the pigeons lost their nesting HABITAT. Perhaps even more damaging to their numbers was the fact that the pigeons were hunted for food. Once killed, birds were packed in barrels and sold in city markets for food.

The HUNTING of passenger pigeons continued throughout most of the 1800s. Birds not killed for food were killed for sport. By the late 1800s, the population was reduced to only several thousand birds. In the 1890s, several states passed laws to protect the passenger pigeon, but it was too late. Once the population became small, predation, disease, low birth rate, and other environmental factors forced the species to extinction.

Although passenger pigeons are now gone, they are still regarded as one of the key sparks that ignited the CONSERVATION movement in North America. In Wisconsin's Wyalusing State Park, a monument to the passenger pigeon reads: "This species became extinct through the avarice and thoughtlessness of man." [*See also* ENDANGERED SPECIES; ENDANGERED SPECIES ACT; MIGRATION; MIGRATORY BIRD TREATY; and NATIONAL WILDLIFE REFUGE.]

Pathogen

Any agent capable of producing disease. Pathogens include disease-causing organisms such as BACTERIA, FUNGI, and protists.

Each pathogen has a characteristic method of transfer. Some are transferred in water, air, or food. Others are transmitted by INSECTS and other organisms. Some, including the pathogens that cause the common cold, the flu, and strep throat, are transferred by direct human contact.

Pathogens are most harmful to humans when they get into the drinking-water supply. Cholera, typhoid, hepatitis, and schistosomi-

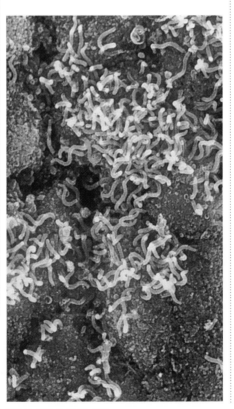

◆ The light-colored, curved bacteria is a pathogen that causes cholera.

asis are diseases caused by waterborne pathogens. These diseases are not a major problem in the United States and other developed nations where public water supplies are routinely checked for their presence. In less developed countries, however, sanitation systems are sometimes inadequate. People may use the same water for drinking, cooking, and bathing. Waterborne pathogens can cause serious problems in these nations.

In 1991, a cholera epidemic struck Peru and rapidly spread to the rest of South and Central America. This epidemic killed nearly 2,000 people the first year. Studies showed the epidemic resulted from poor sanitation and hygiene. [*See also* CHLORINATION; CLEAN WATER ACT; FUNGICIDE; INSECTICIDE; RODENTICIDE; SEWAGE; SEWAGE TREATMENT PLANT; WASTEWATER; WATER POLLUTION; WATER QUALITY STANDARDS; and WATER TREATMENT.]

Patrick, Ruth (1907–)

Pioneer in the use of biological information to assess WATER POLLUTION. Ruth Patrick's research career focused on ALGAE. She was especially interested in diatoms, single-celled golden-brown algae common in almost every water ECOSYSTEM. Patrick is also well-known for her work in limnology, the scientific study of fresh water.

Her work with Charles Reimer on naming freshwater diatoms, stands out in aquatic biology.

Patrick's work with algae and the biological conditions of streams led her to propose in 1948 that biological measures were more effective than chemical measures in assessing water pollution. For more than 50 years, she has led efforts to understand the effects of pollutants on streams and rivers. Her work includes numerous books on the condition of fresh waters.

As curator of limnology at the Academy of Natural Sciences in Philadelphia, Patrick assembled many teams of scientists to study the world's rivers. Her studies took her to Mexico in 1947 with the American Philosophical Society expedition, to the Amazon Basin in 1955 with the Catherwood Expedition, and to almost every major river in the United States.

Patrick is well-known for inventing the *diatometer*, a device used to collect algae on glass slides. Slides colonized at polluted places in a river showed how much contamination they had when compared to unpolluted places. Patrick's work has been rewarded in many ways. She was elected to the National Academy of Sciences and was presented with the John and Alice Tyler Award in ECOLOGY. When Patrick won the Tyler Award in 1975, the $150,000 prize was greater than the amount of that year's Nobel prizes. Patrick also served on the Board of Directors of DuPont, and she has presided over several scientific societies. [*See also* ALGAL BLOOM; BIOME; CARSON, RACHEL LOUISE; HYDROLOGY; and SURFACE WATER.]

PCBs

▌A group of synthetic chemical compounds that cause environmental and human health problems. Before the 1970s, polychloronated biphenyls (PCBs) were used in a variety of products, including electrical devices, paints, inks, PLASTICS, and in coatings for paper, wood, metals, and concrete.

After scientists discovered the harmful effects of PCBs, many nations passed laws to control the use, production, and disposal of these chemicals. The United States banned PCBs in the late 1970s. However, PCBs are still a problem for two reasons. First, they continue to be released from industrial equipment that was made before the ban. Second, PCBs are extremely toxic substances that do not break down quickly. Thus, they are likely to stay in the environment for years to come.

Scientists are not entirely sure how PCBs are released from products into the ATMOSPHERE. However, once in the atmosphere, they can

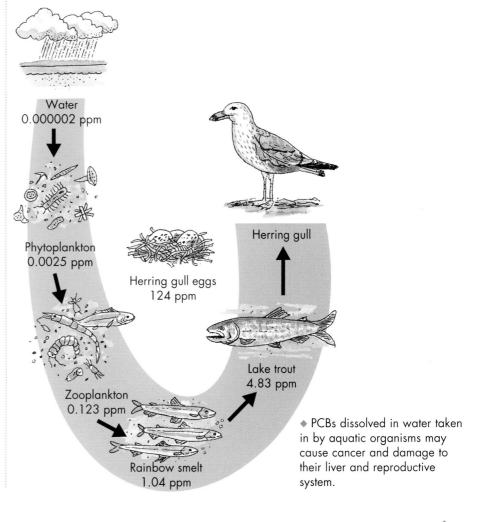

Water
0.000002 ppm

Phytoplankton
0.0025 ppm

Zooplankton
0.123 ppm

Rainbow smelt
1.04 ppm

Herring gull eggs
124 ppm

Herring gull

Lake trout
4.83 ppm

◆ PCBs dissolved in water taken in by aquatic organisms may cause cancer and damage to their liver and reproductive system.

be carried by air currents to different places. PCBs can also enter the environment through SEWAGE, leaking LANDFILLS, and waste dumping.

PCBs cause the most environmental damage when they get into OCEANS, rivers, lakes, and other bodies of water. After an aquatic animal eats PCBs, the chemicals are stored for long periods of time in the animal's body tissues. When that animal is eaten by another animal, the PCBs are passed along to the second animal. As they are passed from one organism to the next, the concentration of PCB s in the animal's tissues increases. This situation, known as BIOACCUMULATION, allows the chemicals to be passed along a FOOD CHAIN in increasingly larger concentrations.

When humans eat FISH or other aquatic organisms contaminated with PCBs, the chemicals can be passed along to people. In humans, PCBs have been linked to CANCER, reproductive disorders, and liver damage. [*See also* BIODEGRADABLE; HAZARDOUS WASTE; MARINE POLLUTION; OCEAN DUMPING; PRIMARY POLLUTION; and WATER POLLUTION.]

Peat

▶A compacted mass of dried plant matter formed by the partial DECOMPOSITION of mosses and other PLANTS. Peat is the first stage in the formation of COAL.

Peat forms in watery areas such as swamps and bogs. The water in these areas has little OXYGEN available for DECOMPOSERS such as BACTERIA and FUNGI, organisms that readily feed on and break down dead matter. When oxygen is not available, the amount of matter decomposed is reduced.

Plant matter that is not decomposed or only partly decomposed builds up on the floor of a swamp or bog. New layers of decomposed plants collect on top of old layers. Over time, the plant matter at the bottom layer is squeezed together, or compacted, by the weight of the matter in the upper layers. In this process, water and air are also squeezed from the spaces between the plant matter. Pressure produced by the weight of the upper layers of matter increases temperature in the lower layers. Over time, the squeezing and increased temperatures cause the partly decayed plant matter in the lower layers to dry and harden, forming peat.

CLIMATE also plays a role in peat formation. Plant material breaks down rapidly in tropical areas, where temperatures are high. Temperate regions have moderate temperatures and relatively small amounts of rainfall. In temperate climates, huge amounts of peat may form in areas having many lakes. As plants in lakes die, they settle to the lake bottom. These organisms form a layer of plant matter. As additional layers of plant matter pile up, the size of the lake and depth of the water decreases. The lake thus changes into a swamp or a bog. In time, trees begin to grow, adding roots and stems to the top layer of plant matter, and *Sphagnum* moss grows on top of this layer. The moss lives on the nutrients in the water and thrives atop the peat

◆ In Ireland and the Netherlands, peat is a common fuel.

formed from the lower levels of plant debris. Over time, the layer of brown moss can grow to a depth of 20 feet (6 meters). Such a formation is called a *peat bog*.

Because it forms from plant matter, peat is rich in CARBON. This trait makes peat useful as both a fertilizer and a fuel. Peat is often dried and burned as a fuel for heating and cooking. It is commonly used for these purposes in the Netherlands and Ireland. [*See also*

AIR POLLUTION; BIOMASS; FOSSIL FUELS; and HYDROCARBONS.]

Pest Control

▌The act of regulating unwanted organisms. A pest is any unwanted organism that occurs in an area.

Pests can be a nuisance in homes and gardens; however, they are a much bigger problem to farmers.

Common agricultural pests include rodents, weeds, BACTERIA, VIRUSES, FUNGI, and INSECTS. Each year, insects alone destroy about 10% of the crops grown in the United States. Today, pests are controlled in a variety of ways, including the use of chemical PESTICIDES, BIOLOGICAL CONTROLS, and INTEGRATED PEST MANAGEMENT (IPM) techniques.

THE USE OF PESTICIDES

Why are pests such a problem to farmers? In the wild, several plant SPECIES usually grow together in an area. This often helps limit the damage done by pests. For instance, one plant might attract insects that kill the pests of neighboring PLANTS. When humans plant large fields containing only one type of plant, they are disturbing this natural type of predator-prey relationship. As a result, single-crop planting, or MONOCULTURE, allows pest populations to thrive because fewer of their natural enemies are present. At the same time, the pest population is provided with a suitable HABITAT that includes an abundance of food.

Sulfur and certain plant chemicals, such as nicotine, have been used to control pests for centuries.

◆ Grasshoppers that severely damage plants are pests that cause problems for farmers.

However, widespread use of synthetic pesticides is fairly recent. Since the 1950s, worldwide use of pesticides has increased rapidly, and its growth is expected to continue. Today, about 2,000,000 tons (1,800,000 metric tons) of pesticides are used each year around the world. That is approximately 1 pound (2.2 kilograms) for each person on Earth.

In the United States, farm pesticide use increased from 145,000 tons (130,000 metric tons) in 1964 to nearly 250,000 tons (225,000 metric tons) in 1990, an increase of nearly 60%. Much of the increase results from the many new synthetic pesticides available to farmers. These pesticides include HERBICIDES, which kill weeds; RODENTICIDES, which kill rodents; FUNGICIDES, which kill fungi; and INSECTICIDES, which kill insects. The majority of these pesticides are used on only four crops: corn, wheat, soybeans, and cotton. About 20% are applied to lawns, gardens, golf courses, parks, and other lands.

DRAWBACKS TO PESTICIDE USE

Worldwide use of pesticides has increased food production. However, there are many drawbacks to the use of pesticides. Chemical pesticides are not only **toxic** to the pests they target but also may be toxic to people and WILDLIFE. Pesticides contaminate many food products on which they are used. If ingested, these pesticides can cause illness in people. Pesticides can also enter water supplies through RUNOFF from farms, gardens, and lawns.

Effects on Human Health

Many pesticides cause people to get sick. In fact, about 50,000 cases of direct pesticide poisoning are reported in the United States each year. Perhaps more serious are the long-term effects of pesticides. Studies have linked pesticides with increased risks for CANCER, birth defects, and many other illnesses.

Effects on Wildlife

The effects of pesticides on WILDLIFE are especially serious because it sometimes takes several years for the chemicals to break down. When pesticides enter the ENVIRONMENT, they can be absorbed and stored inside the bodies of some organisms. The chemicals can then be passed along an ECOSYSTEM'S FOOD CHAIN as one organism eats another. One such long-lasting pesticide is the chemical dichlorodiphenyl trichloroethane (DDT). DDT was a commonly used pesticide in agriculture until some scientists began studying its effects on

◆ The corn earworm, the immature form of a moth, is one of the worst pests in the United States.

wildlife. When it was found that DDT caused the thinning of eggshells in BALD EAGLES and other birds, the United States banned its use in 1972.

Resistance to Pesticides

Another important drawback to the use of pesticides is that pest populations sometimes develop a **resistance** to the chemicals if they are used long enough. Through the process of NATURAL SELECTION, about 500 insect species have developed resistance to pesticides since the 1940s. Resistance to pesticides is also common among rodents, weeds, and fungi. When organisms develop a resistance to certain chemicals, farmers have to apply the pesticides in stronger concentrations or use other more toxic chemicals. Resistance to pesticides also requires that more time and money is spent developing new and more effective chemicals.

ALTERNATIVES TO PESTICIDES

Pesticides are not the only way to control pests. Because of consumer demand for safer food products, many farmers are now turning to IPM techniques that do not rely on harmful pesticides. IPM is a system of pest control that uses a combination of methods for controlling pests naturally without using chemicals.

Biological Control

Biological control is an important part of IPM that involves the use of living organisms to control pests. For instance, one biological control method makes use of the national enemies of pest populations. This method was first used in 1888 to control an insect called the cottony cushion scale, which was destroying citrus crops in California. A natural predator of the scale insects, Australian ladybird beetles were released among the citrus plants. Gradually, the scale insects were

◆ Scale insects may kill a plant by sucking out the sap. California citrus growers once feared losing their trees because of the cottony cushion scale.

◆ Mexican bean beetles may weaken or kill plants by eating their leaves.

eliminated and the citrus crop was saved.

Another popular technique among organic farmers is the use of biological pesticides as alternatives to synthetic pesticides. Biological pesticides are sprays containing microscopic PATHOGENS, such as bacteria, viruses, protists, and fungi. These natural pesticides attack and destroy pests causing them to get sick. Other common organic pesticides are the natural chemicals obtained from some plants, and the use of **parasitic** wasps that invade and destroy the offspring of insect pests.

Physical and Cultural Controls

Physical and cultural control techniques are pest control methods that do not kill pests. Common physical controls are dust barriers, nets, and sticky traps that physically block pests from reaching areas of the plant where they can do damage.

Cultural controls are gardening pesticides that help reduce pest problems by keeping plants healthy. Healthy plants are better able to resist damage by pests and disease than weaker plants. Properly spacing plants and maintaining moist and fertile SOILS are common techniques for keeping plants healthy and free of disease. [*See also* AGRICULTURAL POLLUTION; AGROECOLOGY; BIOACCUMULATION; BIOLOGICAL COMMUNITY; CARCINOGEN; CARSON, RACHEL LOUISE; ECOSYSTEM; EVOLUTION; FEDERAL INSECTICIDE, FUNGICIDE, AND RODENTICIDE ACT (FIFRA); FOOD WEB; HEALTH AND DISEASE; ORGANIC FARMING; PARASITISM; PLANT PATHOLOGY; and SUSTAINABLE AGRICULTURE.]

The remaining entries beginning with P are found in Volume 5.